D0153840

TOWN PLANNING INTO THE 21ST CENTURY

It is nearly 100 years since Ebenezer Howard wrote his famous and influential book *Tomorrow – a peaceful path to real reform*, later to be published as *Garden Cities of Tomorrow*, which was destined to become one of the main influences upon the emerging town planning profession in Britain. During the years of this century, and particularly in the post-war period, town planning became established as the main policy mechanism for dealing with questions of land use and development, the environment and the 'quality' of life in Britain.

However, as we move to the new millennium, it is becoming clear that new approaches and new ways of working will be required if countries such as Britain are effectively to address the new environmental agenda of sustainable development outlined at the Earth Summit in Rio in 1992.

Town Planning into the 21st Century brings together authors who between them have wide experience of the processes of land use and environmental planning. The book's chapters offer a series of insights into the planning process and prescriptions for change that will be required as we move into the twenty-first century.

Contributors: Cliff Hague, Peter Hall, Susan Owens, Eric Reade, Richard Cowell, Bob Colenutt, Yvonne Rydin.

Andrew Blowers is Professor of Social Sciences (Planning) at the Open University; **Bob Evans** is Head of Geography at South Bank University.

FLORIDA STATE
UNIVERSITY LIBRARIES

MAY 31 2001

TALLAHASSEE, FLORIDA

TOWN PLANNING INTO THE 21st CENTURY

Edited by

ANDREW BLOWERS and
BOB EVANS

LONDON AND NEW YORK

HT
169
.G7
T69
1997

First published 1997
by Routledge
11 New Fetter Lane, London EC4P 4EE

Simultaneously published in the USA and Canada
by Routledge
29 West 35th Street, New York, NY 10001

Selection and editorial matter © 1997 Andrew Blowers and Bob Evans
Individual chapters © 1997 Individual contributors

Typeset in Garamond by
Solidus (Bristol) Limited
Printed and bound in Great Britain by
Hartnolls Ltd, Bodmin, Cornwall.

All rights reserved. No part of this book may be reprinted or reproduced or utilized in any form or by any electronic, mechanical, or other means, now known or hereafter invented, including photocopying and recording, or in any information storage or retrieval system, without permission in writing from the publishers.

British Library Cataloguing in Publication Data
A catalogue record for this book is available from the British Library

Library of Congress Cataloging in Publication Data
A catalogue record for this book has been requested

ISBN 0-415-10525-0 (hbk)
0-415-10526-9 (pbk)

CONTENTS

CONTRIBUTORS

Andrew Blowers is Professor of Social Sciences (Planning) at the Open University. His teaching and research are in the fields of environmental planning, policy and politics, with special interest in the politics of sustainable development and radioactive waste. Among the books he has written are *The Limits to Power*, *Something in the Air* and *The International Politics of Nuclear Waste* (as co-author); he has also edited *Planning for a Sustainable Environment*. His most recent work has been as co-editor and author of a three-volume series on *Environmental Policy in an International Context*. A former Dean and Pro-Vice Chancellor of the Open University and former Vice-Chairman of the Town and Country Planning Association, he is also a member of the government's Radioactive Waste Management Advisory Committee and has served as an elected county councillor in Bedfordshire since 1973.

Bob Colenutt is Head of Urban Regeneration at the London Borough of Haringey. He worked as a planner and researcher for community groups in North Southwark and in Docklands between 1972 and 1984, and after a year with the GLC was Head of the Docklands Consultative Committee Support Unit between 1986 and 1995. He was a local councillor at the London Borough of Lambeth between 1986 and 1990 and for three of those years was chair of the Planning Committee. He is co-author of *The Property Machine* (1975) and has written widely on Docklands, development in London and on community-led regeneration. He is an adviser to the Labour Party City 20/20 Inquiry into Urban Policy.

Richard Cowell is Research Fellow at the Department of City and Regional Planning, University of Wales, Cardiff. After completing his Ph.D. on the topic of sustainability, planning and environmental compensation at the Department of Geography, Cambridge University, he has conducted research on various aspects of environmental politics, policy and planning. Present projects include environmental management and compensation, theories of sustainable development, and the regulation of the minerals industry.

Bob Evans is Head of Geography and Housing at South Bank University, London. He is co-author (with Julian Agyeman) of *Local Environmental Policies and Strategies* (Longman, 1994) and is co-editor (also with Julian Agyeman) of the journal, *Local Environment*. He has worked as a town planner in the public, private and voluntary sectors and is author of *Experts and Environmental Planning* (Avebury, 1995) and co-author (with Susan Buckingham-Hatfield) of *Environmental Planning and Sustainability* (John Wiley, 1996).

Cliff Hague is a professor in the School of Planning and Housing at Edinburgh College of Art/Heriot-Watt University. He was the President of the Royal Town Planning Institute in 1996. He has been involved in teaching professional planners in the Czech Republic, Hungary and Pakistan as well as in the UK, and has also published work on planning in China. He has worked as a tutor with the Open University and has been a consultant to them on courses such as 'Restructuring Britain' and 'Environment'.

Peter Hall is Professor of Planning at the Bartlett School of Architecture and Planning, University College London, and Professor Emeritus of City and Regional Planning at the University of California at Berkeley. From 1991 to 1994 he was Special Adviser on Strategic Planning to the Secretary of State for the Environment, with special reference to issues of London and South East regional planning including the East Thames Corridor and the Channel Tunnel Rail Link. He is chair of the Town and Country Planning Association. He is author of some thirty books on planning and related topics, including *Cities of Tomorrow* (1988); *London 2001* (1989); *The Rise of the Gunbelt* (with A. Markusen, S. Campbell and S. Deitrick, 1991), *Technopoles of the World* (with M. Castells, 1994) and *Cities in Civilization* (forthcoming 1997).

Susan Owens is a lecturer in Geography at the University of Cambridge and a Fellow of Newnham College. She has a long-standing interest in land use, environment and sustainability, and has written several books and numerous papers on these and other environmental issues. She has held a fellowship under the ESRC's Global Environmental Change Programme for research on land-use planning and environmental change, and has also conducted research for the European Commission, the Department of the Environment, OECD and environmental NGOs. She was a Special Adviser to the Royal Commission on Environmental Pollution during its Transport study and is a member of the Countryside Commission and of the UK Round Table on Sustainable Development.

Eric Reade was, until early retirement in 1983, Lecturer in Town and Country Planning, University of Manchester. He fled the country to live in Sweden in 1983, but presently mistakenly returned. He claims to be the only planner of his generation who has not become a consultant, and regards the present-day tendencies both to rely upon consultants and to reduce the work of local government to an endless series of separate projects based on competitive bids for funding, as pernicious. He longs for a return to autonomous local authorities, financed by local taxation and able to engage in proper research and sound long-term planning.

Yvonne Rydin is Senior Lecturer in the Department of Geography at the London School of Economics and Political Science. She is the author of *The British Planning System* (Macmillan, 1993) and co-author, with T. Brindley and G. Stoker of *Remaking Planning* (2nd edn 1996, Routledge). During 1994–5 she was the co-ordinator of the RICS Environmental Research Programme; the results of this programme were published under her editorship as *The Environmental Impacts of Land and Property Management* (Wiley, 1996). Her most recent work is on discourses of the environmental agenda (Myerson and Rydin, *The Language of Environment*, UCL Press, 1996) and the urban policy process (published in various articles).

PREFACE

It is conventional to talk of recent history in terms of decades, as if the 1970s were marked off from the 1980s, and the 1990s reflected a new set of changes. This tendency is likely to be all the more pronounced as we enter a new millennium which offers the opportunity both for retrospective pronouncements on a supposedly passing era and for speculative analysis of the prospects for the unfolding years to come. To an extent this periodicity masks the continuities of processes and the cycles of changes that are not so bound by regular demarcation of time. But, recognising that any point in time may represent an ending, a beginning and a continuing, the dawn of a new century (and especially a new millennium) is commonly accepted as a time to reflect on past, present and future. This book is a reflection on planning as we enter the new millennium.

For planning, such stock-taking is especially appropriate. In the first place, planning is a practical activity assessing past trends, making projections and setting out the constraints and opportunities for the future development of our environment. Beyond that, planning, in its broadest sense, is also about visions, the imagination of what the environment could (perhaps should) be like. This vision, or social purpose of planning, has existed much longer than planning as a professional occupation or governmental activity. Indeed, the first great visionary text of planning, Ebenezer Howard's *Garden Cities of Tomorrow* was published almost a century ago, coincidentally at the dawn of a new century. It proved more than mere vision since some of its ideas were put into practice, demonstrating the fusion of vision and practice that has always been present (if often subdued) in planning.

These two faces of planning – the one as a regulatory, governmental and 'professional' activity, the other as a purposive, ideological and analytical programme – constitute the core of the debates in this book. The shifting emphasis in planning practice over the last three decades or so is traced from its emphasis on comprehensive, strategic approaches founded on governmental intervention in the public interest, through a period where market interests became the dominating feature to a present period where public/private partnership is emphasised. Paralleling and influencing these shifts have been changes in the techniques of planning, an earlier fascination with rationality, modelling and systems giving way to a softer focus on planning as an enabling activity facilitating change, leading on to the contemporary concern with the wider issues of sustainability.

The fundamental purposes of planning, so strong in the post-war period, have been largely subdued during the past few decades as planning has retreated to a politically neutral (some might say, neutered) non-ideological role. To an extent, planning may have found its purpose once again as sustainable development becomes the rhetorical by-word for policy making across a range of government policies and in the private sector.

There remain very divergent views of what planning is or should be and they are reflected in this volume, written by some of the leading planning thinkers in Britain today. Some of the chapters, those by Evans (ch. 1), Evans and Rydin (ch. 4) and Hall (ch. 7) reflect on what has happened in planning over these decades in terms of its practice, ideology and achievement. Other writers consider how planning should respond to the contemporary challenge of sustainable development. In this vein, Cowell and Owens (ch. 2) take a pragmatic view of the impact of constraints on planning's prospects while Blowers (ch. 3) envisages a central role for planning within the wider realm of environmental management. Two of the chapters, by Reade (ch. 5) and Colenutt (ch. 6), argue for planning to abandon its pretensions and urge a resurgence of its concern with social justice and community development.

The last two chapters consider British planning in its broadest context, in terms of international comparisons (Hague, ch. 8) and in terms of the relationship between society and the environment (Blowers, ch. 9). As a whole, the book presents the debates about the role and purpose of planning as it goes into a new millennium, irrelevant as that milestone may prove to be.

Most of the chapters were presented originally at a conference organised by the Planning and Environment Research Group (PERG) of the Institute of British Geographers held at South Bank University. The editors would like to thank all the contributors for their co-operation and forbearance as successive drafts were undertaken. Gratitude is also due to those who have supported the project in various ways, including the members of the PERG committee, and to Michele Marsh and Nicola Hallas who provided secretarial support. The result may not change the planning world but it should cause us to think about it, and it is upon thought that change ultimately rests.

Andrew Blowers
Bob Evans
February 1997

1

FROM TOWN PLANNING TO ENVIRONMENTAL PLANNING

Bob Evans

Town planning has been part of British public life for nearly a century. From roots in social reformism emerged the profession and practice of town and country planning which has exerted a singular influence upon the built and natural environments of Britain. The undoubted achievements and triumphs of post-Second World War town and country planning – the New Towns programme, the designation of the National Parks, the introduction of a national system of land-use control – all have to be set against the reality that the brave new world implicit in much of early town planning idealism has failed to materialise.

Poverty, racism, unemployment, urban decay and environmental degradation still characterise many parts of late twentieth-century Britain, and although it might be objected that 'town planning' could never have been expected to address, let alone solve, these apparently intractable problems, it is quite clear that many of the late nineteenth-century 'founding fathers' of town planning had precisely these long-term social purposes in mind.

However, it is not the task of this introductory chapter to review the development of town planning in Britain throughout this century – this has been done admirably elsewhere (for example Cherry, 1974; Rydin, 1993; Ward, 1994). Similarly, it is not necessary to rehearse the extensive criticisms of planning and planners central to the work of Reade (1987) and others. Instead, this chapter examines town planning and the processes of land-use policy in the light of current social, economic and political circumstances, in order to assess how these processes of public policy can and should change to meet the challenges and demands of the new century and to set a context for the subsequent chapters.

The chapter focuses particularly on four issues. First, it examines the concepts of town planning, land-use policy and environmental planning in order to specify the character and nature of these potentially conflicting designations. Whilst 'classical town planning' is now largely defunct in Britain, any developing 'environmental planning' must be a rather different phenomenon. This leads, second, to a consideration of the new environmental agenda and, specifically, the policy goal of sustainable development.

Although there may be similarities between the long-term goals of the early town-planning movement and the objective of environmental sustainability, this new agenda implies a qualitatively new public-policy objective which is unlikely to be secured by traditional approaches.

Third, there is the issue of land policy and the perennial question of land ownership and 'betterment'. As Ebenezer Howard emphasised nearly a hundred years ago, these matters are of pivotal importance to questions of land use and planning and there is a need now to rework these debates again in a new context. Finally, there are questions of equity and democratisation, themes which have permeated British town planning for nearly a hundred years and which are integral to current debates over sustainable development.

TOWN PLANNING, ENVIRONMENTAL PLANNING AND LAND-USE PLANNING

British town planning emerged out of particular cultural, political and economic circumstances as a form of land-use control which is quite different to the systems that have emerged elsewhere. Town planning in Britain is a state policy *process* for allocating land use and deciding on development proposals, and, as such, it is an activity similar to the systems for land-use management and control common in other industrial and post-industrial societies. What makes British town planning distinctive, however, is first its *professionalism*, and second its long-term affiliation to some notion of *reform*.

The creation of the Town Planning Institute (TPI) in 1913 marked a turning point in the development of town planning in Britain. Up until that point the embryonic activity of town planning was principally characterised by reform. The town-planning 'movement' in the first decade of the twentieth century had the characteristics of a social movement, a series of organisations and individuals dedicated to securing societal reform. The principal organisation was, of course, the Garden Cities and Town Planning Association (now the Town and Country Planning Association), which was a pressure group dedicated to pursuing the ideas of Ebenezer Howard. Other important players were the Fabians, the Sociological Society and the National Housing Reform Council. In contrast, by the time of the passing of the 1909 Housing and Town Planning Act, there were only four men (women were not admitted to the TPI until 1928) practising as town planners (Hague, 1984).

The formation of the TPI was to change all this. The other, established, professions – architects, engineers, surveyors – laid claim to town-planning knowledge and expertise, and it was the unwillingness of these existing professions to cede responsibility to another that in large part allowed the creation of an independent Institute and, by implication, an independent profession. Once established, the TPI was careful to keep its distance from the Garden Cities Association, which it considered as militant, propagandist and 'political'. For its own part, the TPI sought to establish itself as technical and non-political – in short it pursued the classic strategies of professionalisation from its inception. By 1920 an examination system had been created to restrict entry into the Institute and the first university department of Town Planning and Civic Design had been established at Liverpool. 'Planning' began to be established as a legitimate, independent and technical activity, distinct from the existing professions.

The 'professionalisation' of town planning has been reviewed elsewhere (e.g. Reade, 1987; Evans, 1993) but a few points bear re-emphasis. It is quite clear that, in terms of theoretical knowledge, there really is no such thing as the 'art and science' of town planning. Whilst it is possible to point to the 'knowledge' that underpins medicine, civil engineering or law, such theoretical backing is absent in the case of town planning. Reade makes this point succinctly:

> If the market [in land] were in fact to be rendered largely inoperative, as would have occurred if the 1947 Act's 100 per cent recoupment of betterment had continued in force ... we would then be brought up sharply against what is probably the central intellectual weakness of the planning profession; in this situation the pattern of land use would *have* to be determined administratively, but 'planning' as a body of knowledge seems to offer no theoretically grounded technical criteria on the basis of which this might be done.
>
> (Reade, 1987, p. 29)

It might be objected that town planning is a practical discipline, within which theoretical knowledge is less important than day-to-day skills, but this is equally problematic. In a survey conducted nearly two decades ago, Healey and Underwood concluded that although the planners they studied had a knowledge of government policy and its implementation, few had any identifiable skills, for example in quantitative techniques or design. This led them to conclude that town planners have succeeded in acquiring financial and occupational status on the basis of a variety of ideals rather than expertise or skills (Healey and Underwood, 1978). A more recent survey has shown that even town planners themselves, when asked, are unable to specify precisely what skills they possess (Evans, 1995b).

'Ideals' are of singular importance in the development of British planning, and in particular it is the assumption that town planning is inherently a 'good thing', and that it is, by implication, a reforming or progressive activity operating in the interests of 'the community' as a whole, which has permeated the profession since its inception. A comparatively recent description of the purpose and role of planning conveys the essence of this:

> It [town planning] deals with the problems of contemporary urbanisation, not only remedying malfunction, but also creating the conditions for harmonious living. (Health, beauty and convenience have long been standing objectives in securing model cities). It deals with the allocation of land for stated purposes; it seeks to relate economic planning to the physical structuring of cities; and it aims to enhance living conditions for the community as a whole.
>
> (Cherry, 1981, p. 4)

It is this ambitious and idealistic assumption that a common interest in a better living environment can be achieved through planning that has strongly influenced planners' perceptions and practice. But ideals and reality have seldom coincided. More often than not, the outcomes of town-planning activity have been socially regressive, benefiting land and property owners and the educated and articulate middle class rather than the 'community' as a whole (Hall *et al.*, 1973).

These arguments about town planning are not new and neither have they been refuted in any substantial way. The purpose in briefly mentioning them is simply to reinforce the point that what might be called 'classical town planning' has run out of steam. The term represents a set of ideas and ideals which were genuinely radical in their time, but, in the context of late twentieth-century Britain, the concept of town planning has little or no validity. Classical town planning is perhaps best defined in this quote from Lewis Keeble's 1950s town-planning textbook. Planning is:

> the art and science of ordering the use of land and the character and siting of buildings and communication routes so as to secure the maximum practicable degree of economy, convenience and beauty.
>
> (Keeble, 1952, p. 9)

Throughout the 1940s and 1950s and well into the 1960s, there was widespread acceptance of this 'art and science'. Indeed, during the period of post-war reconstruction and after, town planning did have considerable physical impact as ideas were put into practice in the form of urban redevelopment, planned population dispersal, rural development strategies, neighbourhood units, green belts, new towns, national parks and other spatial policies. But, as time wore on and political ideologies shifted policy to a market-based stance, so it became clear that planning lacked not merely the powers, but also the expertise and theoretical knowledge necessary to tackle the complex environmental problems facing contemporary society.

The regular calls for town planning to 'return to its roots', or for a new planning consensus (Ward, 1994), are really missing the point. Land-use policy is simply a public-policy process or mechanism, in principle no different to tax collection or waste management, through which public or government policy is enacted. To conflate the policy *process* with policy *ends* is to represent the political as the technical. As Reade (1987) has pointed out, this confusion not only results in poor policy, it is also fundamentally undemocratic.

Given these criticisms, how might town planning be transformed to meet the challenges of the new century? The Town and Country Planning Association have called for a new approach which they term 'environmental planning' (Blowers, 1993a). The reasoning behind this approach is more fully outlined in a later chapter of this book but the essence of the argument is that since the 'environment' cannot be compartmentalised, then neither can policy responses designed to deal with it. Environmental planning therefore, is conceived as an integrated and holistic approach to the environment that transcends traditional departmental and professional boundaries, and is directed towards securing the long-term goal of environmental sustainability. Land-use policy is but one element of environmental planning, along with energy policy, waste management, water resource management, pollution control and so on. Land use is clearly of great importance, but it is not, as it were, *primus inter pares*, and no one professional or occupational group may legitimately claim control over the whole process.

Notwithstanding the difficulties that arise from adopting some notion of sustainability as a policy goal, the argument for a more integrative policy approach to the environment is overwhelming. The underlying theme of all recent environmental policy initiatives, including the UNCED Rio Agenda 21 agreement and the European Union's Fifth Environmental Action Programme, 'Towards Sustainability', is the need for policy mechanisms that reflect the complex and interdependent character of environmental problems. Even the British government's recent national strategy for sustainability (HMSO, 1994a) no longer refers to town planning, preferring instead to discuss land-use planning, and development in town and country within a wider environmental context.

So, what of 'land-use policy'? Given the argument so far, it should be clear that land-use policy is simply an area of state policy-making, albeit an important one. It should not in itself be imbued with particular notions of reform, nor should one occupational group be able to claim control of it. Land-use policy is a policy process or instrument which may be used by governments for a variety of ends, and these ends are determined through the political process, parliamentary or otherwise. Thus, it is folly to 'believe in planning' as one recent past president of the Royal Town Planning Institute exhorted – how is it possible to believe in a process as opposed to ends? By the same token, there cannot be an objective category of 'good planning', as is often implied by professional planners, since this can only be judged in the context of politics – 'good' for whom? These arguments are returned to in Chapter 4.

This focus upon the processes of planning and land-use policy, rather than a concern with policy outcomes, might be viewed as irrelevant or even obsessional. However, it *is* important since the outcomes of policy are, in large part, a consequence of how that policy is framed, organised and implemented. If Britain is to develop a pattern of land use appropriate to the needs of a more environmentally aware twenty-first century, it will be necessary first to locate land use firmly within a process of environmental planning at national, regional and local levels, and second to move land-use policy away from its current focus upon the regulation of land uses towards the utilisation of a wider array of policy instruments, including taxation. Hence, the need for a move from land-use *planning* to land-use *policy*.

The case for linking taxation policy and land-use regulation is, of course, not new. This principle was a fundamental plank of that most radical piece of legislation, the 1947 Town and Country Planning Act. Following the recommendations of the Uthwatt reports, the Act not only nationalised development rights, it also nationalised development value, recognising that a national system of land-use planning could only operate if there was a land-use policy that permitted control over land markets. This theme will be returned to below and is discussed further by Reade in Chapter 5.

SUSTAINABILITY

The discussion so far has centred upon policy mechanisms and processes, but what of the goals? It is quite clear that some conception of sustainability – or sustainable development – has now become the formal goal of environmental policy, including land-use policy, in Britain, Europe and the world in the sense that agreed policy documents have specifically stated that this is so. The agreement to Agenda 21 at the Rio Earth Summit and the subsequent work of the Commission for Sustainable Development; the European Union's fifth environmental action programme (CEC, 1992); the UK government's *Sustainable Development: The UK Strategy* (HMSO, 1994a); and the many local sustainability plans currently being drawn up by local authorities throughout the world – these and other environmental policy programmes accept the goal of sustainability. Moreover, the call for an integrated, holistic approach to environmental planning is a process that is specifically directed at securing environmental sustainability.

As Jacobs and Stott (1992) point out, the terms 'sustainability' and 'sustainable development' tend to be used interchangeably, whereas they represent distinct approaches. They argue that 'sustainable development' is not simply about the environment since it incorporates other indicators of human welfare such as incomes and their distribution, jobs, health, housing, crime levels and so on. Conversely, they argue, 'sustainability' implies a much stronger commitment to the environment above other factors, principally economic growth. This division between an environmentally 'weak' sustainable development and an environmentally 'strong' sustainability is one that reflects the divisions that have existed throughout the development of environmental politics.

For obvious reasons, most local, national and international policy documents are, implicitly or explicitly, concerned with sustainable development. However, most adopt an unproblematic view of sustainable development, and usually this means the ritual incantation of the famous Brundtland definition – that 'sustainable development is development that meets the needs of the present without compromising the ability of future generations to meet their own needs' or some derivative of this (WCED, 1987, p. 49).

In some cases there is a recognition of the difficulties and complexities inherent in this notion, but these are usually brushed aside – the most famous example being John Major's Foreword to *Sustainable Development: The UK Strategy*, where he states: 'Sustainable development is difficult to define. But the goal of sustainable development can guide future strategy' (HMSO, 1994a, p. 3).

Although it is conventional to describe sustainability as a contested concept, it would probably be more accurate to emphasise its capacity to gloss over differences. As Myerson and Rydin (1996a) note, it is difficult to object to sustainability. Clearly, the concept has the capacity to span a wide range of political positions, and it is the very ambiguity of the term that makes it so attractive. By the same token, much of the current debate over the nature and applicability of sustainability indicators is fraught with difficulty precisely because of the wide range of potential interpretations of what sustainability might mean.

These definitional wrangles are in themselves largely unproductive, and the search for an all-embracing and widely accepted definition of what sustainability or sustainable development might mean (beyond a broad agreement to the Brundtland definition) is, consequently, fruitless. However, this is not to reject sustainability or sustainable development as legitimate social goals. On the contrary, a public and informed involvement with the goal of sustainability will increasingly be an essential component of a modern polity. Thus, rather than review here the disparate and often competing interpretations of what sustainability or sustainable development might be, it is more useful to examine the consequences of the adoption of sustainability as a goal of public policy.

It has to be recognised that sustainability is a qualitatively different policy goal to those that have traditionally characterised central and local government in Britain, and for that matter elsewhere. Unlike other goals, sustainability is long term and all-embracing (at least in principle), whereas most other areas of public policy are transitional and specific. Thus, sustainability is a very different public-policy goal from say, 'low inflation', 'full employment', 'more social housing' or 'lower public expenditure'. These goals are specific and easily understandable (at least in superficial terms), and they are quantifiable in that measurable targets are comparatively easily established. Sustainability, however, does not obviously share any of these attributes. Apart from the difficulties of definition and therefore also measurement, sustainability as policy deals in time horizons hitherto unimaginable – certainly decades and probably centuries.

All this leads to the suggestion that sustainability as a policy goal has more of the character of such ideas as 'freedom', 'justice' or 'democracy'. These are what might be termed 'over-arching societal values' rather than specific public-policy goals. And as with freedom, justice and democracy, there are differing interpretations and understandings of what sustainability is and how it might be secured. One might also add that with all these concepts there tends to be a high level of rhetoric and ideology and a rather lower level of objective achievement or reality. Sustainability is, therefore, a qualitatively different policy goal and one which governments, at least in Britain, are finding difficult to handle.

The central point is that sustainability is, at its very heart, a political rather than a technical or scientific construct, and the variety of interpretations of the notion reflect this. For this reason, there is unlikely to be a 'universal theory' of sustainability to inform or guide practice, and sustainability cannot be technicised or reduced to a series of indicators or standards, useful and necessary as these aids undoubtedly are.

In the specific context of land-use policy, it has become commonplace to assert that one purpose of planning is to secure sustainable cities, or perhaps a sustainable pattern of land use. The problem here, of course is that there is no agreement as to exactly what either of these states might be. Although the examinations by Elkin *et al.* (1991), Owens (1991), Breheny (1992), and others, have provided key insights into how a more 'green' and energy-conscious land-use policy might operate, it is by no means clear that, for example, dispersed patterns of living are inherently less sustainable than concentrated, higher density settlements.

This apparent absence of certainty should not be a cause for alarm. On the contrary, as was noted above, one major problem with existing land-use policy has been the domination of policy by professional experts who have had a clear interest in representing political questions as technical ones. One response to the 'problem of sustainability' described above is to argue that once the overwhelmingly political character of sustainability is understood, our approaches to it as a public-policy goal can become much clearer. There can be no doubt that politicians and policy-makers will need to seek advice from a wide variety of experts – ecologists, economists, waste-reduction and recycling technologists and so on. But, if environmental planning for sustainability, with land-use policy as a central component, is to be anywhere near effective, the political processes of public debate and controversy, both formal and informal, will need to play a much more significant role than has hitherto been the case.

As political theorists from Rousseau onwards have maintained, an active, politically participatory society is more likely to be successful in securing widespread public support for societal goals, and in these circumstances 'activity' is likely to foster 'stability', since participants see themselves as stakeholders. These arguments have a special salience in the context of sustainability. As Jacobs (1991) and others have so clearly argued, democracy is fundamental to sustainability since it is very clear that an environmentally sustainable society can only be a possibility if large numbers of people abandon existing attitudes and adopt new ones which may not be in their immediate short-term interests. These problematic questions of democracy and the related issue of social equity have a high profile in the current global environmental debate and they have deep roots in the ideology and rhetoric of British town planning.

LAND-USE POLICY AND THE 'BETTERMENT QUESTION'

Ebenezer Howard's most influential book, *Tomorrow: A Peaceful Path to Real Reform* (later retitled *Garden Cities of Tomorrow*) (Howard, 1902), is usually cited as a powerful argument for the creation of new free-standing settlements away from existing towns, and as such it is rightly viewed as a seminal work in the development of British town planning. However, it is important to recognise that Howard's project was based upon a clear understanding that such objectives could only be achieved if 'the community' controlled the land market.

This link between the planning process and the value and ownership of land – often referred to as the betterment or development-value debate – is central to the land-use policy process and has a long and hotly contested history that has been extensively examined elsewhere (e.g. Cullingworth, 1980; Reade, 1987) and does not require repetition here. Instead, aspects of this debate are highlighted in order to emphasise the importance of the public recoupment of development value as part of a revitalised process of environmental planning.

'Development value' is most commonly understood as the difference between the existing use value of a piece of land and its value if it can be converted to a more profitable use. In contemporary Britain this usually refers to the increase in value of a piece of land which is brought about as a consequence of the granting of planning consent for a more profitable land use. Thus, for example, the owner of farmland who secures planning permission to develop it for housing (or whose land is zoned for residential use) usually achieves a significant increase in the value of the land, and it is this difference between the existing use (agriculture) and the more profitable land use (residential) which is usually termed the 'development value'. In this case the increase in the value of the land has occurred because the 'community', through the land-use planning system, has decided that housing is a desirable land use. The increase in value has occurred through public action rather than the action of the land owner. There has been no change in the form or character of the land itself and, for this reason, development value, following John Stuart Mill, is often termed the 'unearned increment'.

Clearly, land values can increase in other, somewhat similar ways. The overall level of economic activity may increase the demand for land or, as a result of infrastructural investment – for example, new roads or airports – land values in particular locations may increase. Nevertheless, in all of these cases the increase has been socially produced through the operation of economic processes (the market) or by specific governmental action, and it is this that underpins the claim that since development value is produced by the community, it should therefore be owned by the community.

The Uthwatt Committee (1942) clearly recognised that if the new post-war system of land-use control was to be effective, it would be essential to link this with mechanisms for taxing betterment. Thus, the 1947 Town and Country Planning Act nationalised both development *rights* and development *value*. Not only was this logical in terms of the arguments outlined above, it was also necessary in order to ensure that the land-use planning system had a sufficient level of control over the market so that it could be effective in actually promoting positive planning schemes rather than simply controlling development. The hope was that the 100 per cent betterment levy would eventually create a situation where all land was traded at existing use value, thus enabling public authorities to assemble land banks for development purposes with all development value reverting to the state. In fact, the charge stifled the market so that sufficient land did not come forward, leading to the dropping of the charge when the Conservatives returned to power.

However, with the abolition of the development charge in 1953 (and a subsequent return to market value as the basis for all state compulsory land acquisition in 1959) the British land-use planning system became what Reade terms 'pseudo-planning, the appearance of planning without the reality. It seems likely that in such a system it will often be the market rather than planning which decides' (1987, p. 23). The post-1953 land-use planning system was to become largely impotent in the face of powerful property interests.

The current land-use planning system in Britain is mainly characterised by what has been termed 'trend planning' (Brindley *et al.*, 1989). In other words, its predominant feature is the tendency to accommodate and support market trends. If sustainability is to be a real policy goal, a much more positive, proactive planning approach will be required which will inevitably need to offer strong opposition to market forces. As Reade (1987) points out, this will only occur when taxation and land-use policy are formulated together with the long-term objective of removing the privileged position that the land market holds over the land-use planning system.

Moreover, sustainability implies a much more careful usage of all non-renewable resources, and a more thoughtful consideration of how land is allocated and used than is currently the case. There is a real need to reduce the pressure for land development which, in times of economic upturn such as the late 1980s, is substantially fuelled by the opportunity that land owners have to make often quite large windfall gains of development value. In addition, the recoupment of betterment, even if this were to be done at a percentage considerably less than the 100 per cent tax of the 1947 Act, could provide a substantial income to fund public investment in environmental programmes.

It is for these reasons that many informed commentators on the British land-use planning system have argued for one or another of the various schemes to recoup development value. The Town and Country Planning Association sees such a process as central to their call for environmental planning (Blowers, 1993a), Reade has argued for Site Value Rating (Reade, 1987), whilst Hall opts for auctioning planning permissions (Hall, 1989). The advantages and disadvantages of the different schemes for recouping development value are discussed elsewhere (Reade, 1987; Evans, 1995a), but the central issue for most of these authors is not so much 'which scheme is best?' but 'which scheme is politically possible?'.

Many established, powerful interests and organisations are implacably opposed to any scheme of betterment taxation, and their position is strengthened by the apparent total failure of the three schemes tried during the last fifty years. The Conservative Party has acted quickly to repeal any Labour legislation seeking to recoup development value. Moreover, the nature of the issue in hand is such that it will never attract widespread public passions – it is not a vote catcher and mobilisation of popular support for this measure will be difficult if not impossible. Nevertheless, despite these difficulties, the Labour Party's recent environmental policy document, *In Trust for Tomorrow*, clearly indicates the Party's intention to ensure community benefit from increases in site value resulting from the granting of non-domestic planning permissions (Labour Party, 1994, p. 46). But, even for the Labour Party, the land question remains a neglected issue and has not featured in its manifesto commitments at recent general elections or in its latest (1996) policy pronouncement.

This matter of betterment lies at the very core of land-use policy and will not go away. There are very substantial political and administrative obstacles to be overcome before any mechanism for recouping development value could be implemented in Britain. Nevertheless, until there is an equitable system that is subservient to a nation-wide process of land-use control and regulation of development, the market will continue to dominate land-use policy. In these circumstances, the prospects for developing a sustainable pattern of land use in Britain appear dim.

EQUITY, DEMOCRACY AND ENVIRONMENTAL PLANNING

As was noted above, the roots of British town planning lie, in major part, in social reform, and as a consequence of this, the twin objectives of greater social equity, and the increased involvement of citizens in the processes of public decision-making have tended to feature regularly in much town-planning practice and professional and academic literature. In the main, the link between planning and questions of equity has been implicit rather than explicit, via an assumption that planning is, unarguably, a socially progressive activity that will improve the conditions of those at the bottom of the socio-economic hierarchy. The precise mechanisms for securing these changes have rarely been specified, but there is often an implicit physical determinism which assumes that improvements in socio-economic conditions arise from changes in physical environments or land use.

During the late 1960s and 1970s British town planners were influential in encouraging the development of participatory structures in planning decision-making. Following the recommendations of the Skeffington Committee (1968), the town and country planning legislation was amended to incorporate a statutory right to public consultation in the process of local and structure plan preparation. Whilst this might be viewed as a comparatively minor development, town planning decision-making is one of the few areas of UK public policy where such measures have emerged, and this was, in part, due to pressure and encouragement from the planning profession. However, as a consequence of the Rio 'Earth Summit' held in 1992, these questions of equity and democratisation have been given a new impetus, and they may now be regarded as an integral part of the national and international 'new environmental agenda'.

The enigmatically titled 'Agenda 21' is in many ways the most influential environmental document ever to have been signed at the international level. It is generally regarded as the most important agreement to have emerged from Rio, and it was signed by most of the attending national governments, some 150 in number. Agenda 21 is a 500-page document that sets out how both developed and underdeveloped countries can work towards sustainable development, and it specifies the actions that will need to be taken by the world community if development is to be reconciled with environmental concerns. In addition to requiring a reduction in the usage of energy and raw materials, and of pollution and waste, Agenda 21 also represents a call to share wealth, opportunities and responsibilities more fairly between North and South and between rich and poor, both nationally and internationally. In this sense, Agenda 21 is, in Levett's words, 'profoundly democratic and egalitarian in outlook' (Levett, 1993).

The document not only emphasises the need to adopt policies and strategies that meet the needs of disadvantaged groups, it also stresses the importance of encouraging such groups (women, youth, indigenous peoples) to participate in decision-making and in the implementation of policy. The concept of 'capacity building', therefore, has a prominent place in Agenda 21. It emphasises the need for people and organisations to develop the capacity to undertake and implement policies that will contribute to sustainable development.

As a consequence of the incorporation of conceptions of greater social equity and democratisation into Agenda 21, these notions have secured a high profile within environmental policy during the last few years. Although it must be recognised that there is currently a substantial gap between the rhetoric of greater equality, empowerment and democracy, and the emergence of action determined to secure these objectives, it is clear that for many writers, politicians and policy-makers, increased equity and democratisation are integral to sustainability.

The rationale for this position rests upon a belief that sustainability inevitably implies a sharing of common futures and fates and hence some degree of perceived equity. In many cases, at least in the prosperous North, sustainability will probably mean the adoption of policies that will threaten current lifestyles and patterns of consumption, and these are only likely to be accepted if it seems probable that all members of society will be affected equally. Moreover, as Jacobs points out, if sustainability is collectively agreed and enforced, it may actually win wider approval than if it is expected to rely upon individual choice and action (Jacobs, 1991, p. 128).

The nub of this problem is that there tends to be a very close association between economic exploitation and environmental exploitation and, as Blowers points out later in this book, if sustainability is to be a politically realistic goal, then inequalities in environment and development must be addressed by the richer nations of the world. However, evidence from the 1995 Berlin Summit on Climate Change suggests that the richer nations, the oil producers and the rapidly developing countries of the South, are mostly reluctant either to enter into agreements that will reduce energy consumption and greenhouse-gas production, or to transfer to the South the wealth and technology needed to reduce their long-term impact upon the environment.

Thus, although greater social equity and the democratisation of decision-making may be widely regarded by environmentalists as essential components of sustainability and environmental planning – locally, nationally and internationally – it is very clear that those social groups and nations possessing political and economic power are unlikely to surrender this on the basis of some kind of environmental altruism. Given the exponential nature of the growth of environmental problems, these attitudes may change quite rapidly, as short-term self-interest is displaced by an awareness of the longer-term consequences of environmental inaction.

During the next decade, the profession and practice of land-use planning in Britain will need to change and adapt to meet the demands of a new century and a growing concern with environment issues. The nature and form of these changes, and the way in which they are implemented, will clearly have considerable consequences for the future natural and built environments of Britain. The remaining chapters of this book are intended as a contribution to the debate over the future of land-use planning in Britain as we approach the new century. They may also help to revive the campaigning spirit of the town-planning reformers of nearly a hundred years ago so that we, in turn, can seek to meet the environmental challenges of the new millennium.

2

SUSTAINABILITY: THE NEW CHALLENGE

Richard Cowell and Susan Owens

INTRODUCTION

Land use and environmental change are connected in fundamentally important ways, and land-use planning is one of the oldest instruments of environmental protection. However, the perception of 'environment' and its treatment in planning policy and practice have changed significantly over time and have broadened considerably as we approach the twenty-first century. Healey and Shaw (1994a) have traced the evolution of environmental themes from the welfarist-utilitarianism of the 1940s to the current preoccupation with sustainable development. Whether this latter theme represents a new paradigm or a recasting of old conflicts is a significant question for planning theory and practice in the coming decade.

The potential contribution of land-use planning to sustainable development has already been recognised at all levels from the global to the local: by the United Nations (UN 1992), the European Commission (CEC 1992) and by national, regional and local authorities (for example, HMSO 1994a, LGMB 1993). In Britain, the role of land-use planning has been promoted by government policy statements as well as by planning policy guidance (see, for example, HMSO 1994a, DOE 1992b, 1992c, DOE and Department of Transport 1994). Planning authorities are urged to take environmental considerations comprehensively, and consistently, into account in development plans, and to integrate environmental concerns into all planning policies. The profession has responded with enthusiasm, but also some perplexity, to this challenge.

An important effect of this new theme has been an extension of the remit of the planning system beyond its traditional, essentially local, concerns with land use and amenity to encompass the environment in a much wider sense: 'The boundaries between land-use planning, environmental planning and sustainability ... are now blurred' (Cullingworth and Nadin 1994: 137, Blowers, Chapter 3 of this volume). Newer concerns are exemplified by global warming, depletion of non-renewable resources and the cumulative impact of development decisions on biodiversity. These links between land use in specific localities and a broader conception of 'environment' up to the global scale have important implications for the future development and purpose of the planning system.

The visible and politicised arena of land-use change is an important testing ground for sustainable development because it demands the translation of abstract principles into operational policies and decisions. This act of embedding interpretations of sustainability in specific environments will raise complex and difficult issues extending beyond the locality in question and beyond the present remit of the land-use planning system. If sustainability is to form a basis for planning in the twenty-first century, certain (long-standing) dilemmas will need to be confronted with renewed urgency. We address three such issues in this chapter: questions of value and subjectivity; the relationship between land-use planning and ideologies of need and efficiency; and the limits to 'impact managerialism'.

Our discussion centres on interpretations of sustainability deriving from concepts of 'environmental capital', and their implications for planning policy and practice. These implications are illustrated by an increasingly contentious area of planning – the provision for aggregates extraction. We look particularly at an inquiry into a minerals local plan in Berkshire and some of its wider ramifications. Emerging theories and practice point to important issues for planning in the twenty-first century, which we identify in our concluding section.

INTERPRETATIONS OF SUSTAINABILITY

The concept of sustainability has complex origins and diverse meanings (see, for example, Pezzey 1989, Redclift 1990). Some commentators distinguish it from 'sustainable development', suggesting that social dimensions are integral to the latter, but we have not maintained such a distinction in this chapter. Planners themselves have been grappling with different definitions and operational principles, as reflected in numerous conferences, debates in the literature and reports from professional bodies (County Planning Officers' Society 1993, Blowers 1993a, Welbank 1993). Engagement with these issues extends to the Regional Planning Conferences and to many statutory and non-governmental organisations (NGOs) with a keen interest in planning processes and outcomes (for example, Countryside Commission *et al.* 1993, English Nature 1994, Jacobs 1993, RSPB 1993).

An increasingly influential interpretation of sustainability is that of maintaining 'environmental capital' (Pearce *et al.* 1989, Pearce and Turner 1990, Daly and Cobb 1989, Jacobs 1991). This idea draws heavily on economic metaphors and is based on two key principles. The first is that justice between generations means bequeathing to the next generation a stock of 'capital', with its assumed capacity to produce well-being at least equivalent to that enjoyed by the present. The second evokes the precautionary principle of challenging the orthodox assumption that all forms of capital are substitutable. Recognising that some functions of the environment are vital and irreplaceable, it counters the assumption that products of economic growth (human-made capital) provide unproblematic substitutes. Instead, it requires that social and economic activity should be managed at least to conserve such 'critical environmental capital'. Some authors incorporate this rule within a broader model of 'constant environmental capital', which not only protects what is critical but maintains at least the present value of the environmental capital stock. 'Non-critical' capital is not considered inviolable, but any loss or damage must be compensated by some equivalent environmental benefit – a process of 'environmental compensation' (Cowell 1993).

This interpretation is significant because, although 'sustainable development' was never synonymous with unlimited growth, it undoubtedly won broad appeal by giving the impression that economic growth and environmental protection were mutually compatible, if not synergistic (WCED 1987, Weale 1992). However, when interpreted as maintaining environmental capital, sustainability shows a distinct affinity with the older discourse of 'limits' because both reject the notion that biophysical capacities are infinitely elastic. This does not mean that growth is necessarily limited but it does imply that, in order to be sustainable in the long term, the nature of growth must be such that it respects constraints set by the need to maintain critical environmental capital (and in some interpretations the total value of the environmental capital stock) intact. Hence, there are strong links between maintaining environmental capital and the controversial concept of environmental capacity.

IMPLICATIONS FOR PLANNING

These constructions of sustainability have proved attractive to professional planners as well as to others involved in land-use change and development and they are beginning demonstrably to influence planning practice and policies. Whether they do much to modify *outcomes* is an interesting and important question which we can begin to address through empirical work. First, however, it is worth asking what is so compelling about these concepts that they have been greeted with such enthusiasm in planning circles. We suggest three possible explanations. First, concepts of sustainability centred on 'environmental capital' seem to have meaning at a variety of spatial scales and can be interpreted in ways that suit the geography of the planning system: we can conceive of stocks of environmental assets, critical environmental capital and environmental capacities not only at the global or national scale, but in regions and counties, districts, towns and even villages. Defining the state of the environment – as it is and as it 'ought' to be – for one's own planning area has intuitive appeal and fits well with idealised models in which targets are set and policies implemented and monitored. A second, related, explanation is that these interpretations of sustainability offer an apparently coherent framework for conservation and compensation: in particular, the concept of critical environmental capital provides a rationale for something approaching inviolability – the 'trump card' that has hitherto eluded those seeking to protect important features of the natural and cultural environment. Finally, the delineation of environmental capital and its protection into the (distant) future requires planning *writ large* and provides a means of empowerment. It is hardly surprising that after a decade in which some of the basic premises of the planning system came under attack, such an opportunity should be seized with enthusiasm.

Applying principles of sustainability in practice, however, remains problematic. The reasons for this are complex, involving not only problems of interpretation – exactly how concepts of sustainability can be mapped onto particular environmental features – but also difficulties in defending 'sustainable' principles and policies once they have been defined. Although determining what is sustainable is sometimes represented as a quasi-technical exercise, in practice it forces planners to confront competing value systems, including their own. It will inevitably become strongly contested ground; all the more so because of a fundamental incompatibility between 'maintaining environmental capital' and certain principles which have underpinned the functioning of the land-use planning system since its inception.

Prominent among these is the overarching principle of 'balance', with the planning system seen as a pivot on which the benefits of development and conservation should be weighed against each other. The balance has traditionally been weighted by a presumption in favour of development, which places the burden of proof in land-use conflict with those seeking to show that development should *not* proceed. In a sense, the benefits of development are taken as axiomatic but the benefits of protection and conservation have to be demonstrated. Though this burden of proof is shifted in areas protected by national or international designation, there is invariably a clause permitting development for 'imperative reasons of over-riding public interest' (see, for example, DOE 1994b). The reluctance to imply absolute protection is understandable, but in practice such clauses are open to expedient interpretation (MacEwen and MacEwen 1982). Under the new plan-led system, the same broadly favourable presumption applies to development that is in accordance with the plan, a step that shifts attention, conflict and the burden of proof to the plan itself.

Neither the presumption in favour of development nor the ubiquitous requirement to 'balance' different material planning matters sits easily with concepts of environmental capacities, and each of these certainly conflicts with the notion of critical environmental capital. If it is a function of planners 'to identify environmental capacities and prevent them from being breached' (Jacobs 1993), the implication is that there would be a presumption *against* development that breached these capacities, however defined. Furthermore, if critical environmental capital must be handed on intact to future generations, it is effectively removed from 'the arena of trade offs' (Collis *et al.* 1992: 20). While the identification of such constraints must itself involve a broader assessment of priorities, this implies not a case-specific balance sheet but a more fundamental analysis of what we value in the environment and why.

Attempts at defining stocks of environmental capital to be maintained, and at designating some of these assets as 'critical', reveal complexities in the meaning of 'environment' and in the interpretation of what is sustainable which have profound significance for the planning system. The concept of what is sustainable relates more readily to some meanings of 'environment' than to others. Brundtland's influential argument that 'Development cannot subsist upon a deteriorating environmental resource base' (WCED 1987: 37) is most convincing in relation to *material* dimensions of environmental concern such as resource degradation, health and survival (Owens 1994). The needs of both present and future generations can be defined here without great difficulty, even if precise limits to pollution and resource exploitation are always contested. Boundary conditions defined by critical loads and maximum sustainable yields are familiar and have clear antecedents in long-established concepts of prudent resource use. In this context, however, land-use planning has had a relatively restricted remit, though progress with state of the environment reporting, environmental indicators (for example LGMB 1993) and emerging aspects of pollution policy may begin to challenge traditional boundaries.

Environmental capital also embraces *postmaterial* dimensions of environmental concern, involving the amenity, aesthetic and non-instrumental values inhering in both the natural and cultural environments (Owens 1994). So the 'stock' must include habitats, landscapes and cultural assets, and it is frequently argued that a sub-set of these should be regarded as 'critical' (CSERGE 1993, Blowers 1993b). This extension is significant because these are the areas in which the role of the planning system has long been legitimised. But defining what is sustainable here is difficult because it requires an explicit theory of value in fields that generally command less consensus than the sanctity of life and health. Issues of landscape and habitat preservation, for example, raise profound questions about how values are expressed and measured, whether they are widely shared and how to act when they are not. Conservation of species and ecosystems is, of course, linked to material concerns about human health and survival, but this is not always its primary rationale.

There are further important implications. By definition, what is critical must be passed on intact to future generations, whatever the aggregate benefits to the current generation of actions that might modify or destroy it. For this very reason, the definition of critical environmental capital will be strongly contested: and when we are dealing with amenity, aesthetic and intrinsic values, the potential for dispute seems almost infinite.

To summarise, if sustainable development is ultimately about living within our environmental means, it must respect some a priori environmental constraints on economic activity; yet respecting 'environmental capacities' would represent a paradigm shift from forty years of post-war emphasis on 'balance' and the presumption in favour of development. It is difficult to predict the impact of sustainability concepts on the practice and philosophy of land-use planning, but we suggest three broad scenarios for planning in the twenty-first century. In the first, the new ideas have no significant impact on planning that essentially accommodates developments generated by a demand-led economic system: this is business-as-usual. In the second, sustainability provides a vehicle for a significant shift in priorities – a changing of weights in the balance. In the third, sustainability heralds a shift towards an environment-led system – a new paradigm for environmentally constrained land-use change. If we move towards the latter, it is possible to envisage a number of issues dominating policy debates over the next decade and beyond.

1. The first concerns the construction of defensible arguments for protecting any particular function of the environment as 'environmental capital'. Defining capacity constraints in relation to pollution and human health raises complex issues of risk and scientific indeterminacy, yet it may prove easier to conceive of critical loads, for example, than critical landscapes.
2. We might expect that, in combination with the new plan-led system, defining what is sustainable will expose, at an earlier stage in the planning process, many of the conflicts that sustainable development was meant to reconcile. This is because placing parts of the environment into categories – particularly critical status – sets a limit to future growth which effectively pre-empts consideration of individual development proposals on their merits.
3. These issues are intensely political because sustainability constraints feed into and alter current patterns of economic activity (indeed, it is advocated that they should; see, for example, House of Lords Select Committee on Sustainable Development 1995). As a corollary, they challenge entire ideologies of social need; how it is to be defined and met and how it is reconciled with 'needs' for the environment.
4. Prominent in these debates are the various technical discourses of impact management and compensation. Tight defence of environmental capacity is controversial because it claims a priori that certain forms of development cannot be acceptable in a particular place, in advance of any claims by developers about either the benefits of their development, or skills at impact management and habitat creation.
5. Questions also arise about linkages – with other policy instruments and between localities and scales. The planning system clearly has a limited

ability in isolation to effect the economic and other changes necessary to bring the structure of economic activity into line with environmental capacities. There is also a need to consider how tightening environmental capacities in one locality affects communities and environments elsewhere – the question of exporting unsustainability.

These issues seem intractable but must be confronted if we are to make sense of sustainability in real policy contexts. Indeed, they are already being raised. With some variations, these interpretations of sustainability have been seized with vigour by environmental organisations concerned with protecting amenity, the countryside and with nature conservation (Jacobs 1993, RSPB 1993, Collis *et al.* 1992). The challenge they represent is beginning to be recognised with considerable unease by developers. Local planning authorities, too, have begun to show an interest in such 'capacity-led' planning, as we illustrate below with the case of Berkshire's draft replacement minerals local plan. Through empirical analysis of these ideas within actual planning contexts, we might begin to discern which scenario for planning and sustainability is most likely to dominate the next decade.

We turn now to the increasingly contentious issue of aggregates extraction, though many of the arguments apply with equal validity to a wide range of developments where meeting some defined social 'need' demands the consumption of valued environments. It is at just such interfaces that interpretations of sustainability will be tested in the political domain.

AGGREGATES, SUSTAINABILITY AND PLANNING

Background

Broadly speaking, the supply of aggregates in the UK has been governed by a 'predict and provide' philosophy in which projected demand, strongly linked to economic growth, has been equated with a need which should be met. Similar principles have applied to other commodities but, for aggregates, this philosophy has been particularly pronounced. National demand forecasts have been translated by the Department of the Environment (DOE) into Regional Guidelines indicating how provision for the supply of materials should be made to meet anticipated demands. The guidelines then provide a basis for Regional Aggregates Working Parties (RAWPs) to agree allocations for particular counties. These bodies, under review at the time of writing, draw their membership from local authorities, the minerals industry and central government, and have been a key component of aggregates planning since the early 1970s (DOE 1989). Through these mechanisms, the planning system has been seen essentially as an instrument for accommodating national 'need' while minimising environmental disruption.

Though revisions to minerals planning guidance (MPG) in England have left the essentials of this system in place, they do embody some shifts in the thinking that underpins minerals planning (DOE 1994a; MPGs are issued by the DOE to advise local authorities on minerals planning). The relatively modest but significant changes in guidance (MPG6) published in 1994 reflect powerful lobbying during its preparation (Mabey 1993), the growing political difficulties of gaining consents for aggregates extraction, especially in the South East and, significantly, a new priority for environmental considerations. It is acknowledged that 'a gradual change from the present supply approach is called for', to meet objectives 'in a way which is consistent with sustainable development' (DOE 1994a: para. 25). While MPG6 still holds that 'it is essential that the construction industry continues to receive an adequate and steady supply of aggregates so that it can meet the needs of the community and foster economic growth' (para. 9), it also stresses 'the importance of combining economic growth with care for the environment in order to attain sustainable development' (para. 10).

Although long-term projections are still predicated on links between economic activity and demand for aggregates, their purpose has become less clear. They 'inform' the Regional Guidelines, and environmental implications still have to be 'carefully balanced against the *need* for the material' (DOE 1994a: para. 2, emphasis added). But the preparation of development plans 'provides an important opportunity to *test* the practicality and environmental acceptability at the local level of the Guidelines figure' (para. 58, emphasis added). Significantly, 'alternative sources' (to traditional land-won provision) are expected to make an increasing contribution to supply; the aggregates and construction industries are urged to minimise waste and achieve greater efficiency, and targets are set for the use of secondary and recycled material.

Aggregates and sustainability

Before looking at a specific case study, it is worth considering how concepts of sustainability have been applied to the exploitation of non-renewable resources like minerals. Resource depletion has always been an important dimension of environmental concern and, while it is inevitable that non-renewable resources *will* be depleted, it is generally accepted that sustainability in this context means an emphasis on efficiency of extraction and use coupled with the development of substitutes as resources become scarce (see, for example, DOE 1994a). Sustainability requires decreasing the materials intensity of economic activity (materials use per unit of GDP). Demand for aggregates, however, is still strongly linked to economic growth and, indeed, the aggregates intensity of economic activity has been *increasing*.

Broadly speaking, however, the physical availability of aggregates is not an issue of major concern. It is the impact of aggregates extraction on other environmental assets that dominates the debate. As *Sustainable Development: The UK Strategy* puts it:

> Very large quantities of minerals resources exist, sufficient, in many cases, to last far into the foreseeable future. Nevertheless, it is becoming increasingly difficult to find sites that can be worked without damaging the environment to an extent that people find unacceptable.
>
> (HMSO 1994a: 90)

It is in this context that what is 'sustainable' in terms of minerals extraction needs to be defined, yet a dilemma of conflicting needs cannot be avoided. If, as MPG6 maintains, an adequate and steady supply of aggregates is 'essential' (para. 9), the implication is that valued environments will have to be consumed. If, on the other hand, legitimate demands for environmental quality are at least as important, and some environmental functions are critical, the implication is that sustainability constraints must be determined a priori and should be integral to the *definition* of minerals reserves (the portion of the resource base that is available under prevailing economic and technological conditions). Though potential conflicts can be mitigated through environmental assessment, good working practices and high standards of restoration, it remains difficult to reconcile these two positions; a situation complicated by the fact that the demand for minerals is expressed in markets whereas the demand for environment generally is not. Further complexity is added by the spatial/geographical dimension: minerals can only be extracted where they occur; priorities differ within and between different communities and localities; and difficult questions arise about the 'ownership' of environmental assets. All of these issues are illustrated by the case study to which we now turn.

THE BERKSHIRE MINERALS LOCAL PLAN: PLANNING FOR SUSTAINABLE DEVELOPMENT?

Background

In the autumn of 1993 a public inquiry was held into the draft Replacement Minerals Local Plan for Berkshire (the Plan) (RCB 1993). We focus here on the general philosophy of the Plan, which drew heavily on concepts of sustainability, and the challenges to it, particularly those from mineral operators and their trade associations. Berkshire County Council's key argument was that to make provision for aggregates production at the rates indicated in Regional Guidance after 1996 would breach the environmental capacity of the County. The Plan proposes to meet commitments to make provision for 2.5 million tonnes a year of aggregates production up to 1996 and thereafter to reduce this amount by 3 per cent per annum. A total of 458 objections were made to its general policies, and a further 238 to the proposed 'preferred areas' in which it was deemed that mineral operations might be acceptable, subject to normal planning considerations. The inquiry ranged over many issues, but three in particular merit detailed comment because they go to the heart of questions about sustainability: the issue of 'need' and its relation to demand forecasts; the legitimacy of the concept of environmental capacity; and the meaning of 'efficiency' in aggregates production and use. Debate on these issues was vigorous, and had added intensity at a time when publication of the revised MPG6 was pending.

Need

No parties questioned the fact that aggregates are needed or that Berkshire is a significant source of aggregates, the County Council fully recognising 'that Berkshire has a part to play in providing this basic resource for a prosperous national economy' (Babtie 1993: 3; the Babtie Group Ltd is the County's planning consultant). The industry equated need with projected demand as expressed in the regional apportionments, at one extreme holding that the prime aim of the Plan should be 'to ensure that the industry was provided with the quantity and type of mineral that it needed' (Brundell 1994: 5). Most operators, however, accepted some notion of the need for aggregates being 'balanced' against other considerations, with specific applications being treated on their merits. Berkshire, in contrast, maintained that there was 'nothing sacrosanct about the regional apportionment ... in the sense that it sets down a level of provision which has to be met at all costs' (Babtie 1993: 22). According to the County, it was explicit in the agreement reached by mineral planning authorities in the South East that the regional apportionment was *to be tested* in the preparation or amendment of development plans (SERPLAN 1989), and implicit that it might be shown to be untenable after 1996 because of environmental constraints.

The Inspector at the inquiry into the Plan accepted that the preparation of local plans offered an opportunity to test post-1996 production figures, but reaffirmed the central role of regional apportionments (Brundell 1994: 12). The Council, he argued, had given insufficient weight to the need for local land-won aggregates and had moved to a position in which provisions were based on environmental considerations. The implications of this were 'profound':

> if each authority in the South East and beyond adopted the same approach it is likely that severe constraints would be placed on the production of aggregates which have a vital role to play in the national economy.
>
> (Brundell 1994: 13)

The implication is that the need for aggregates, expressed in the regional apportionment, is to all intents and purposes taken as given. 'Testing' is acceptable but the County would have to be able 'to demonstrate very clearly the reasons why it cannot maintain a production of 2.5 mt/year beyond 1996' (Brundell 1994: 13). The Inspector recommended that the relevant policy (Policy 3) be modified to allow for production of sand and gravel at an average of 2.5 million tonnes a year after 1996.

Environmental capacities

An important factor behind the Inspector's recommendation concerning production levels was his rejection of the concept of environmental capacity 'in terms of the definition of specific areas within which mineral working would be acceptable' (Brundell 1994: 13). The industry challenged this concept, denying that 'environmental capacity', 'critical environmental capital', or indeed any concept of strict environmental limits had standing in government planning policy on sustainable development. Despite the County's 'persuasive explanation' of the derivation of environmental capacity and its relationship with the concept of sustainable development, the Inspector broadly concurred with the industry's view, recording that the term is 'not yet in common use and ... plays no part in national policy guidance in general or in relation to minerals in particular' (Brundell 1994: 13).

The root of the objection seems to concern both the weight that Berkshire had placed upon environmental constraints and a more fundamental tension between the necessity for subjective judgement and the restrictiveness of capacity constraints. The Inspector noted that in constructing its environmental capacity, 'some factors had been treated as absolute constraints by the county council even though mineral working would be permitted ... in terms of national policy guidance' (Brundell 1994: 13). He accepted that an element of subjectivity in selecting the preferred areas was inevitable and that the site-selection process adopted by the Council was 'thorough and correct' (p. 27). However, this legitimate subjectivity 'in itself demonstrates that it is impossible to impose absolute limits on acceptability. Values must change over time ... and there will always be differences in judgements about particular areas' (p. 14). Since he did not accept that the constraints in excluded areas were always so severe that they could not be overcome, he concluded that the council had not identified all of the potentially acceptable areas for mineral working within the plan period.

In short, according to the Inspector, since 'environmental capacities' are subjective, they do not provide an appropriate basis for determining production levels, which should continue to be based on the regional apportionment. Presumably, an extension of this reasoning would always make it impossible to use subjective judgements to identify sustainability constraints; since values change, the Council will never be able to identify all sites where mineral working might become acceptable. It is a separate point that improving technologies of extraction and restoration may make working acceptable at more sites in future. As it stood, in the view of the aggregates lobby, the plan pre-empted the site-specific balancing function of development control and played down the role of extraction in creating valuable environments such as amenity lakes or nature reserves. It was suggested by one aggregates company that minerals extraction only borrows land, it does not deprive future generations of anything; tight environmental limits on extraction, on the other hand, would deny them the benefits of modern urban fabric.

Efficient use

By planning unilaterally not to meet its SERPLAN apportionment after 1996, the County was open to accusations of NIMBYism. In an attempt to avoid 'exporting unsustainability', however, the Plan incorporated numerous policies designed to reduce demand for primary land-won local aggregates to make up for the shortfall: by minimising wastage, making the most appropriate use of high-quality materials and increasing the use of recycled and secondary aggregates where appropriate. These are all components of 'environmental efficiency' and are encouraged in the most recent version of MPG6 as important components of sustainable development (DOE 1994a: para. 11).

However, policies designed to husband mineral resources in the County attracted vigorous criticism from minerals interests. British Aggregates Construction Materials Industries (BACMI) thought that the policy could not be justified at all 'since there was no shortage of resources in the county' (Brundell 1994: 10). Berkshire's expectation of using increased amounts of recycled and secondary materials was dismissed as excessively optimistic, and it was pointed out that these substitutes have environmental impacts of their own during transport and processing. Moreover, the minerals lobby was reluctant to accept that current aggregates use was in any way wasteful, revealing a particular market-based definition of efficiency. The job of the industry was to respond 'efficiently' to the parameters of costs and product specifications set by the market and society at large, and if that meant using high-grade materials where a lower grade would suffice, then this was not something that they or the local planning authority could, or should, do anything about. Although the mineral operators and their associations felt that there might be some role for the use of economic instruments to adjust markets at the national scale, they opposed any attempt by local planning authorities to usurp control of the end uses of materials.

Regional decisions in a national context

Most of the fundamental principles of Berkshire's plan were severely criticised by the minerals industry, and received only qualified support in the Inspector's report. In particular, the challenge to demand-led planning was not upheld, which might lend support to the view that the influence of sustainability in the planning system is simply to provide another weight in what remains a balancing process. However, because the revised MPG6, published at the same time as the Inspector's report, set out reduced regional production levels, Berkshire's allocation in absolute terms ultimately fell in line with the plan. Arguably, this reduced allocation itself emerged out of conflict over aggregates extraction in the South East and represents an acknowledgement on the part of central government that environmental capacity constraints – for regions as a whole if not for individual counties – have become a legitimate policy consideration. Significantly, *Sustainable Development: The UK Strategy*, published during the period in which the Inspector was preparing his report, represents the planning system as a means of ensuring that growth takes place 'in a way that respects environmental capacity constraints' (HMSO 1994a, para: 35.4).

If the planning system is to achieve this goal, it will be necessary to take a broader perspective. During the inquiry, neighbouring county councils tacitly accused Berkshire of regional NIMBYism, fearing that Berkshire's restricted environmental capacity risked off-loading further aggregates extraction into their patch (Brundell 1994). In practice, the pressures for increased supplies have been decanted beyond the South East region altogether.

Perhaps the most dramatic response to have emerged over the last twenty years is the concept of the coastal 'superquarry', an extraction site with aggregate reserves of at least 150 million tonnes and an annual output in excess of 5 million tonnes, mostly exported by sea. One such quarry is already in operation at Glensanda on the west coast of the Scottish mainland and, at the time of writing, Redland Aggregates Limited await the outcome of an application for a superquarry at Lingerbay on the sparsely populated Hebridean island of Harris. A public inquiry into this proposal lasted from the autumn of 1994 to the spring of 1995 – Scotland's longest ever planning inquiry. For the Western Isles Islands Council, the prospect of increased employment proved attractive and initially it voted to accept the proposal, placing faith in a series of planning conditions and legal agreements to control quarry design and working practices. Opposition groups and local objectors regarded this decision as insufficiently precautionary and have also taken issue with the underlying need for the development. The contingency of the forecast 'need' for the superquarry, and whether it is sufficient to override the constraints on development in a National Scenic Area, received considerable and detailed attention at the public inquiry. By the end of the inquiry, local public opinion had shifted against the development and the Islands Council withdrew its support during the closing days.

This application illustrates the problems – political, social and environmental – of pursuing sustainability through the planning system largely by means of 'locational accommodation'. By promoting coastal superquarries, the government may have hoped to defuse burgeoning environmental conflicts while side-stepping the need to reform the traditional demand-led approach in the aggregates sector. However, objectors to the Lingerbay proposal challenged the view – implicit in government policy – that the remoteness of Scottish coastal landscapes, and the apparent fragility of local economies, provided opportunities for accommodating new sources of aggregates. To risk drastic alteration of the environment, economy and culture of Harris, rather than seeking to manage national aggregates demand was, to some commentators, highly unsustainable.

CONCLUSIONS

Together, the conceptual framework and emerging empirical evidence suggest that the following issues are likely to dominate attempts to interpret and apply concepts of sustainability within the land-use planning system:

1. The interpretation of sustainability in terms of environmental constraints flies in the face of prevailing ideologies, which dictate that predicted demands should be met. The determination of needs is largely exogenous to the local planning process. In the case of aggregates, for example, although production allocations can be 'tested', ultimately local plans must give considerable weight to accommodating them.

   This begs the question – which can readily be extended to other commodities – of why the 'need' for aggregates is given greater priority than the 'need' for environment. Some interpretations of sustainability imply that the latter must take precedence in the case of 'critical environmental capital'. While some environmental preferences are arguably less important, or 'subjective', this is surely also true of the demand for minerals. Some 24 per cent of aggregates, for example, goes into new road construction (DOE 1994a) which has been widely criticised for being an inappropriate way of solving national transport problems. Much of the supposedly 'neutral' demand for aggregates that planners must accommodate reflects specific ideologies of social need and views of 'the national interest' which might be challenged. Associated with these ideologies are dominant paradigms of 'efficiency' which favour market-led concepts rather than ideas based on 'environmental efficiency' in any form.
2. Any commitment to maintain a certain level of environmental capital must define the 'space' available for economic activity, even if it is not related in a deterministic way to the level of activity that might take

place. The status of critical environmental capital pre-empts, or at least strongly circumscribes, the possibility of a trade-off. This suggests that interpreting sustainability in development plans, far from reducing the level of conflict, will expose fundamental contradictions at an earlier stage in the planning process.

3. Both theory and practice reveal problems in developing defensible justifications for any particular environmental constraint or capacity: this is not only a scientific but a political task, despite the quasi-technical language. The problems are at their most acute in relation to less tangible features of the environment, such as amenity, though they also present familiar difficulties in defining environmental parameters such as safe levels of pollution (Beck 1992, Owens 1994). That judgements are necessarily subjective is not in itself a ground for criticism: this does not mean that they are arbitrary or capricious. However, even if subjective views can be seen as a legitimate part of planning judgement, it is a large step, in current circumstances, to convert these views into capacity constraints, rather than merely material considerations. The problem stems from the fact that any definition of environmental capital places restrictions upon current patterns of economic activity with pervasive effects. In such conflicts, as the Berkshire case illustrates, projections of 'national need' for materials are afforded an 'objective' status that takes precedence over 'subjective' judgements of environmental value, even when the latter have some claims to be rigorous and defensible.
4. The issue of subjectivity is further complicated by competing claims about the extent to which impacts can be acceptably managed, or by the argument that development simply produces new environments which are at least equal in value to those they supplant. An emphasis on impact management also shifts debate from ethical matters of environmental value towards technical issues of site-selection, risk and habitat re-creation techniques. Such issues are political in the widest sense, and connected, because if it proves defensible to claim that an impact can be acceptably managed, this reduces the incentive to re-appraise more fundamental questions about 'need' and supply (see Figure 1).
5. Important issues arise concerning the scale at which sustainability is a meaningful concept. If we accept that there must be elements of subjectivity in defining what is sustainable, there will inevitably be differences of opinion both between and within communities about what constitutes valued environmental capital. Planners are not unfamiliar with these problems, but there are wider issues concerning national (and international) economic activity, reconciling such activity with environmental protection, and the allocation of power to decision makers at different scales. It could be argued, for example, that the geographical shift of aggregates extraction from the South of England to Scottish coastal superquarries constitutes 'exporting unsustainability'. But what if the receiving community does not judge the development to be

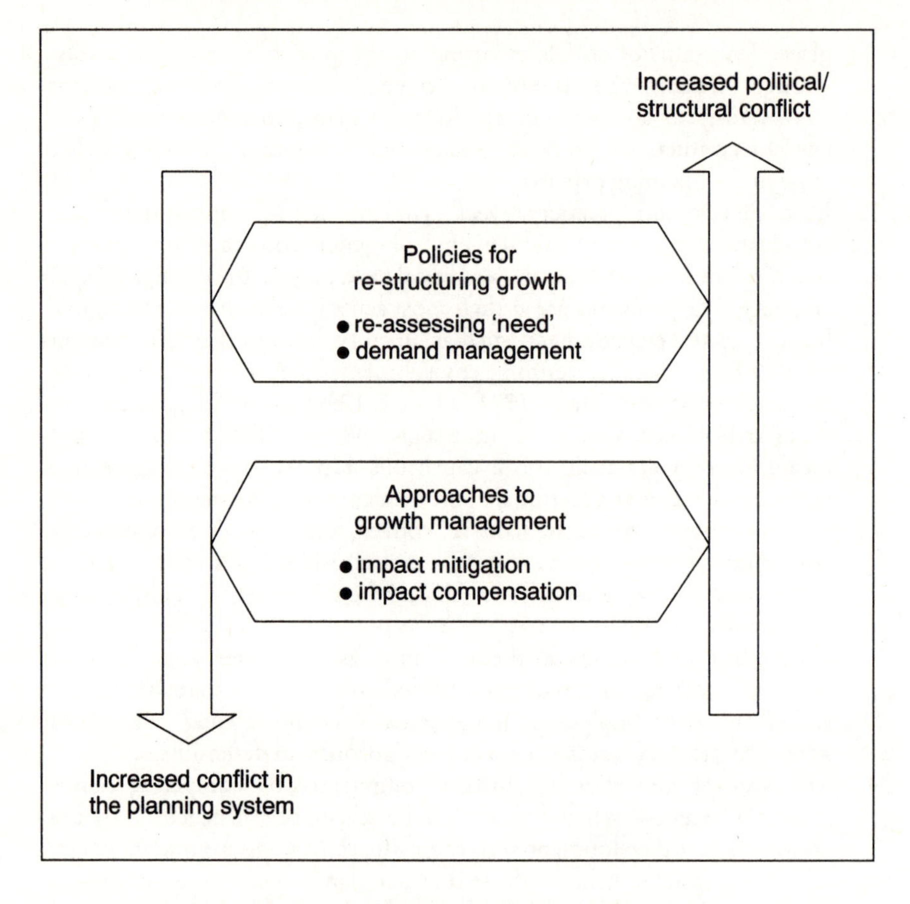

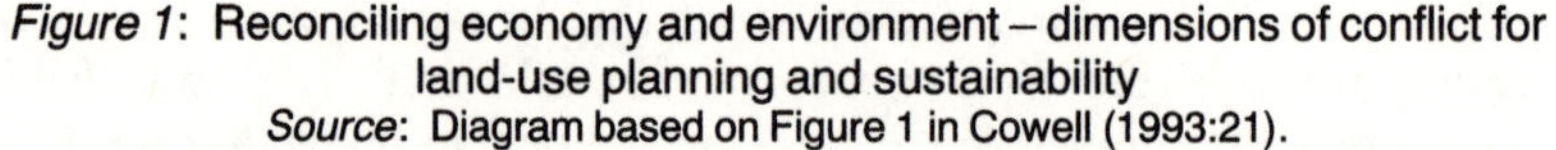
Figure 1: Reconciling economy and environment – dimensions of conflict for land-use planning and sustainability
Source: Diagram based on Figure 1 in Cowell (1993:21).

unsustainable? This raises difficult questions about differentials in economic power, about the 'ownership' of environmental assets, and about the status of 'existence value' – which people assign to environments independently of their use for them – where these apparently conflict with the economic interests of local communities. The Lingerbay inquiry also suggests that conventional representations of economic interests – and alternative ways of achieving development in particular localities – merit careful scrutiny.

Quite how the concept of sustainability will affect planning in the twenty-first century depends on the extent to which these dilemmas are confronted. These are not issues for the planning system alone, however, but have fundamental political implications for wider policy-making arenas. Despite central government's stated support for land-use planning as a contribution to achieving sustainability, our analysis of aggregates planning suggests that 'official' and industry interpretations of sustainable development remain at a significant distance from our third scenario, that sustainability will shortly constitute a new paradigm for environmentally constrained land-use change. So, perhaps partly by default, much depends upon how local planning authorities themselves respond to the new challenge of sustainability, both individually and in concert with other bodies.

ACKNOWLEDGEMENTS

Susan Owens held an Economic and Social Research Council (ESRC) Global Environmental Change Fellowship during 1993–4 to research issues of land-use planning and sustainability. Richard Cowell was an ESRC-sponsored research student until October 1994. The Council for the Protection of Rural England (CPRE) provided support for a further three months' research on minerals planning and sustainability.

3

ENVIRONMENTAL PLANNING FOR SUSTAINABLE DEVELOPMENT

The international context

Andrew Blowers

SUSTAINABLE DEVELOPMENT – A WINDOW OF OPPORTUNITY?

With the ending of the Cold War at the close of the 1980s it appeared that a window of opportunity had opened for dealing with global environmental and development problems. Fears of global Armageddon were replaced, for a while, by heightened concern about environmental security. Arms reduction would create a 'peace dividend' freeing resources that could be be deployed on environmental conservation. This was a time when scientific evidence about the destruction of the ozone layer and the problem of global warming was being publicised. There was also growing evidence of global environmental problems such as deforestation, species destruction and environmental degradation. In the lands of the former Soviet Empire the scale of environmental destruction and pollution was revealed. Yet, in a mood of presumptuous optimism, the rich countries of the West argued that economic growth and environmental conservation must go hand in hand. In short, the triumph of the market economy was an essential precondition for ecological survival.

A way forward also seemed to be to hand in the idea of sustainable development popularised after the publication of the Brundtland Report in 1987. Sustainable development captures, in a single phrase, both the need for economic development and for environmental conservation. It expresses the requirement for social processes to respect the limits of the earth's resources and capacity of its ecosystems. Its ideas have been readily embraced in a series of statements and policies at every level of government and in the private sector (DOE, 1990; HMSO, 1994a; CEC, 1992). It reached its apotheosis at the Rio conference in 1992 dedicated to Environment *and* Development.

By that time, the political energy and priority for the environment was already abating. Just when the size of the problems and the costs of dealing with them were being recognised so the opportunity began to slip as recession engendered a more introspective concern with national economic priorities. In any event, the brave new world order was proving to be turbulent and unstable with regional conflicts in the Middle East, endemic civil wars in parts of Africa, the former Yugoslavia and elsewhere and dangers of nuclear proliferation as the Soviet Union collapsed and North Korea flexed its muscles in the Far East. Instead of a sense of global unity and purpose in which environmental protection might flourish, the world of the early 1990s presented a very different prospect, fragmented and fractious, riven by regional groupings as nation-states tried to grapple with immediate economic and political problems. It appeared that the nation-state, the only universal political organisation with the potential capacity to enact and implement environmental legislation, was suffering 'from a crisis of legitimacy and a crisis of capacity' (Thomas, 1993, p. 1). The political conditions for achieving policies for sustainable development did not look nearly as propitious in the early 1990s as they had seemed to be at the end of the previous decade.

In any case the rhetoric about sustainable development had obscured the reality of the task. There had emerged a passive reassurance that action is both necessary and possible. There was the pervasive belief that change could be achieved through consensus solutions based on commonly agreed goals that could be delivered by the present economic system.

> Sustainable development does not mean having less economic development: on the contrary, a healthy economy is better able to generate the resources to meet people's needs, and new investment and environmental improvement often go hand in hand.
>
> (HMSO, 1994a, p. 7)

Yet the meaning and process of sustainable development remained elusive and vague. In particular the relationship between environment and development was unclear. The Brundtland Report emphasised the need for fundamental changes 'in righting an international economic system that increases rather than decreases inequality' (WCED, 1987, p. 122).

There has been a tendency either to ignore or to downplay the inherent tension and conflicts between environment and development and an unwillingness to confront the political constraints that may render the whole process stillborn (Blowers, 1993b). In particular, the primacy accorded to market forces gives priority to short-run economic goals and neglects the long-term needs of environmental sustainability. In contrast to this, sustainable development requires the adoption of long-term goals, the means to achieve them and a system of monitoring to ensure that targets are being met. In short, environmental resources must be protected and environmental degradation be prevented by 'planning' for sustainability.

It may seem perverse to argue the case for 'planning', a process that has come to be pejoratively associated with bureaucracy, interference, control and inefficiency in an age when the market is ideologically supreme. But planning and markets are not necessarily in conflict or incompatible. As Jacobs indicates, planning 'stands in contrast to the operation of market forces, but it does not preclude the existence of markets' (1991, p. 125).

This chapter takes up the case for a system of 'environmental planning'. It argues the necessity for a comprehensive, integrated and strategic approach to environmental management. The rest of the chapter is in four sections. In the first, the case for intervention through planning is presented. The second section examines the political constraints and opportunities for planning in terms of the three Es – externalities, evidence and equality. The third section outlines the goals and processes of planning for sustainable development. The final section of the chapter considers the changing social and political context and the implications for the environment. Throughout the chapter the UK is portrayed within the broader international context.

THE CASE FOR INTERVENTION

In a world dominated by the ideology of free-market capitalism in which multi-national corporations exercise 'awe-inspiring power' (Lang and Hines, 1993, p. 37) it is not surprising that a market-based solution to the problem of sustainable development is widely advocated. Industry promotes this in its own interest and governments feel comfortable with the notion that economic growth and sustainability are apparently entirely compatible. The belief in market solutions has led to the vigorous search for economic instruments as a means of valuing environmental assets, giving signals to consumers and producers that will lead to resource conservation and lower pollution (Pearce *et al.*, 1989; Cairncross, 1995). Politically, the aim is to avoid regulation where possible. This is explicitly stated in the UK's Strategy for Sustainable Development, 'The Government's general policy is to reduce and simplify regulations wherever appropriate. And in environmental policy, the commitment is to make use of economic instruments where possible, rather than regulation' (HMSO, 1994a, p. 34). The problem with the approach is that it relies on monetary evaluations of environmental assets which rest on contemporary (usually Western) individual preferences and on relative scarcity. This cannot be reasonably applied to different cultural preferences nor to the preferences of future generations and must ignore or discount environmental assets that have yet to be discovered (in the tropical forests for instance). Above all it is impossible to put a value on absolute scarcity, the possibility of ecosystems becoming overburdened or the global commons becoming overpolluted and posing a threat to human survival. Intervention is needed that gives priority to the public or common interest and to the needs of future generations. There needs to be a shift from private to public interest.

At present society is moving in entirely the opposite direction with disastrous environmental consequences. At the international level, the economic interests of individual nation-states take precedence over the common environmental interest. Moreover, the situation is complicated by the conflicts among diverse national interests. Tropical forests are a classic expression of this with conflicts between Northern interests in timber, pharmaceuticals and other products; Southern national interests in gaining economic advantage from the use of the forests; indigenous people's rights; and the common interest in the forests as a source of biodiversity and as carbon sinks. The development of sustainable practices to preserve the forests involves restraint and sacrifice of present economic interests, very hard to achieve in the context of the modern global economy.

In the UK the long period of privatisation, deregulation, cuts in public expenditure and attacks on local government have resulted in a 'democratic deficit' – a dispersal of power to unelected quangos and business interests – and have led to unsustainable developments. A good illustration of this is the largely unrestrained development on the edge of cities of competing superstores serviced by heavy lorries and serving car-borne consumers while the indigent rely on a declining public-transport system to shop in decaying town centres. This process of 'Tescoisation' or 'Sainsburyfication' has only been arrested in the name of sustainable development once the damage has been done.

At both national and international levels, then, there must be an assertion of the public interest. The use of resources and the level of pollution must be controlled in the common interest in sustainability. This can only be achieved by restraint on the private market and a commitment to greater intervention. In particular, intervention will be needed to control property rights in order to prevent overexploitation of resources or environmental degradation. All this should be achieved within a democratic, participative framework. But a shift from private to public property rights will not be achieved without great conflict between the interests of the disadvantaged and of the powerful, who already enjoy a disproportionate share of wealth. This conflict exists at both local and global levels.

The obvious point needs to be made that the environmental preoccupations of individual countries or communities are not separate but are integral and interdependent elements of a broader context. Increasingly, the linkages between local issues and universal processes are being recognised. Development permitted in one area has impacts elsewhere; refusal of development in one area simply transfers the problem elsewhere. Indeed, it is decisions taken at the local level, or by companies or at the level of the nation-state that collectively create and distribute global environmental problems. Agenda 21 recognises that local authorities have a crucial role in that they:

> construct, operate and maintain economic, social and environmental infrastructure, oversee planning processes, establish local environmen-

tal policies and regulations, and assist in implementing national and subnational environmental policies.

(UN, 1992, p. 233)

Consequently, planning for sustainable development must recognise the interdependence between action taken locally and its impact globally. It is within this context that the constraints on policy making can now be examined in terms of the three Es: externalities, evidence and equality.

CONSTRAINTS AND OPPORTUNITIES

Externalities – the problem of responsibility

Strictly speaking in a physical, scientific sense, a global environmental process is one that potentially can affect the whole world. But, 'global' has also a political meaning. It refers to those processes of interaction (co-operation or conflict) between nation-states brought about by environmental problems that transcend political boundaries. Thus, 'globalization refers to the multiplicity of linkages, and interconnections between the states and societies which make up the modern world system' (McGrew, 1992, p. 23). This is a rather inclusive definition which appears to incorporate a whole range of problems that are international in scale but not, strictly speaking, necessarily global. Whatever definition is used it is clear that global environmental problems have anthropogenic causes and trans-boundary impacts, and therefore can only be dealt with through supranational political processes. *The increasing scale of environmental problems means that national economic interests must be subordinated to the broader common interest in sustainability.* In particular, international political action will be necessary to deal with the *externalities* imposed by one state upon others or on the global commons. In addition there is the problem of risks being passed on from one generation to the future.

Under present economic arrangements income and expenditure are measured within companies or countries and defined by boundaries of ownership or sovereignty. The purpose of companies is to maximise profit and the purpose of countries is to increase national wealth. As far as the environment is concerned two problems arise. One is that environmental resources tend either to be undervalued or altogether ignored, regarded as free goods in the calculation of income and expenditure. As a result, some non-renewable resources may become exhausted and renewable resources (including the atmosphere) may eventually become so degraded by pollution as to become irrecoverable. It is the failure to identify these irreversible environmental consequences of contemporary patterns of economic development that has led to the argument that environmental costs must be included in calculations of economic performance (Pearce *et al.*, 1989; Daly and Cobb, 1989).

If this were done there would be an economic incentive to change production patterns and to employ appropriate technology to prevent pollution, to reduce the volumes of waste and to re-use and recycle materials. But valuing environmental assets is not straightforward. There are bound to be different valuations placed upon environmental assets by different societies at different times. Since any value will impose costs there is certain to be conflict of interests over what is acceptable. And, even if agreement on valuations could be found, there would still be the problem of devising appropriate ways of imposing such costs.

The second problem is that environmental costs are not confined within a company or national territory but may be imposed on third parties or on other nations. The unregulated market is not organised to deal with these externalities or to allocate them in sustainable ways. Left to itself, the contemporary global economic system will lead to global environmental deterioration. Hardin (1968) provides an insight into this tendency. The original analogy of the Tragedy of the Commons demonstrated the tendency for the commons to become overgrazed by the collective impact of herdsmen acting in their own interests. Putting the analogy in the context of the global commons (atmosphere, oceans, land), Hardin concludes that it will be rational for companies (or nation-states) acting in their own interests to continue to deplete resources and increase pollution well beyond the regenerative capacity of the earth's ecosystems. They will desist only when resource costs or pollution burdens begin to impact on their profits or wealth. This appears to be a good example of a Catch-22 for policy makers. When you can't see the problem but could act to prevent it, it isn't politically realistic to do so. When you do see it and can act you can't because it is too late to take effective action. As Hardin puts it:

> The rational man finds that his share of the wastes he discharges into the commons is less than the cost of purifying his wastes before he releases them. Since this is true for everyone, we are locked into a system of 'fouling our own nest', so long as we behave only as independent, rational, free-enterprisers.
>
> (Hardin, 1968, p. 1245)

The Tragedy oversimplifies the situation. National self-interest is not easily definable and certainly is not a singular concept persisting unwaveringly through time; rather, there may be a multiplicity of competing and conflicting interests that the state must try to satisfy at any one time. Nor do countries have free access to the commons – instead there are bilateral, multilateral and global agreements that seek to prevent externalities or to compensate for them and which also attempt to protect the global commons from general deterioration. Indeed, a major element of the North–South conflict arises from the attempts on the part of the South to get the North to cut back its polluting activities and to provide sufficient resources to enable Southern countries to achieve more environmentally sustainable policies.

The Tragedy, however, does underline three important political implications of externalities. One is to identify the conflict between individual, private or national interest in immediate economic gains and the longer-term public or common interest in environmental security and survival. A second is the tendency for environmental impacts to manifest themselves at a point when deterioration may already have become irreversible (or remediable only at extremely high cost). And the third implication is that to avoid global damage, agreement needs to be reached between states to deal with the problem of externalities. Such agreement means securing some form of intervention that will override the free market or limit the sovereignty of states.

Evidence – the problem of proof

Concern about global environmental deterioration has revived neo-Malthusian predictions that current exponential trends in resource use and pollution will go beyond sustainable limits and result in 'global collapse' (Meadows *et al.*, 1992). Lovelock concludes that 'if the world is made unfit by what we do, there is the probability of a change in regime to one that will be better for life but not necessarily for us' (1989, p. 178). There is a well-entrenched scientific consensus that global warming poses a palpable hreat to the survival of ecosystems on a world scale within the next two generations. At least at the rhetorical level, there is a growing recognition that sustainability requires that precautionary action must be taken, if necessary in advance of conclusive scientific evidence, to avoid the possibility of irreversible effects. In the words of the Rio Declaration, 'Where there are threats of serious damage, lack of full scientific certainty shall not be used as a reason for postponing cost-effective measures to prevent environmental degradation' (Principle 15). A crucial element in the political process will be the quality and interpretation of the *evidence* of global environmental change.

The concept of sustainable development incorporates a principle – sustainability – which recognises that there are ultimate limits to the capacity of the earth's natural systems to provide the resources and to maintain the environmental conditions necessary to sustain a population of a particular species. If current trends continue it appears that, sooner or later, those limits will be breached as a result of contemporary development, which both destroys resources and pollutes ecosystems. But the evidence of environmental change gathered by scientists is often incomplete, uncertain, conflicting and, consequently, contestable.

Scientific evidence presents general trends rather than precise forecasts. Scientists across the world are in some agreement that the process of global environmental degradation now in train may become irreversible if corrective action is not taken. In very general terms the deterioration is documented in a variety of sources. For example, it is calculated that desertification advances at an annual rate of 6 million ha. (WCED, 1987, p. 128); tropical forests are depleted at a rate of 20 million ha. per year (Holmberg *et al*., 1991, p. 116); at present rates of exploitation, global reserves of coal will be exhausted within 250 years, oil within 30 years and gas within 50 years (RIVM, 1989, p. 22) and so on.

Many of the estimates are very vague indeed. This is especially the case with global impacts that could be irreversible. The scientific experts on the International Panel on Climate Change calculated that average global temperatures are rising at around 0.3°C per decade. But there was a range of uncertainty of between 0.2°C and 0.5°C implying sea-level rises of around 6 cm per decade with a range of 3 cm to 10 cm. Predictions of specific impacts on different parts of the globe are even more tentative. For example, the degree of warming is predicted to be 50 to 100 per cent above the global mean in the high northern latitudes in winter. Little more can be said than that 'global warming could have important detrimental effects on agriculture, forestry, natural ecosystems, water resources, human settlements and coastal protection' (HMSO, 1994b, p. 66).

With biodiversity we are plunged into even greater uncertainty. Only about 1.4 to 1.8 million species have been documented out of a total that could be anything from 10 million to 100 million. Estimates of destruction suggest that between 4 per cent and 50 per cent of species may become extinct between 1980 and 2000 mainly as a result of tropical deforestation. In the UK the loss of species has been catalogued with, for instance, 95 per cent of lowland meadows, 50–60 per cent of lowland heath and 30 per cent of upland heath and grassland lost since 1949. In the case of the vast bulk of species world wide that may become extinct, all that can reasonably be said is that the potential value is simply unknown and therefore incalculable. In addition, existing plants and animals possess evolutionary potential giving rise to new species in the future (HMSO, 1994c). It is perhaps not surprising that the loss of biological diversity 'is one environmental issue that many environmentalists now believe surpasses all others in terms of long-term global impacts' (Barnes, 1996, p. 217).

The uncertainty of much of the evidence is compounded by problems of interpretation. There is the problem of establishing the causes and effects of environmental impacts. For instance, there was the long-running attempt by the UK to evade responsibility for acid-rain damage on Nordic lakes and forests on the grounds that the precise links between sources of SO_2 emissions and impacts such as eutrophication and die-back of trees were impossible to determine. At a global level, the problems of relating sources of greenhouse gases to global and local impacts over long time periods in conditions of scientific uncertainty make it impossible to assign any clear responsibility for causes.

The scientific problems of imperfect evidence and difficulty in establishing cause and effect are compounded by the interpretation of the evidence to suit political interests. It was in the interest of the UK to avoid the costs of acid-rain abatement. At a global level, individual nation-states will strive to minimise the blame attaching to them for global warming, ozone depletion and other environmental changes. The evidence for environmental change is uncertain and leaves the way open for differential interpretations of the significance of the problems and the allocation of responsibility for them. In these circumstances conflicting interests are likely to make it difficult to apply the precautionary principle. At the heart of the political problem of sustainability is the problem of social inequality.

Equality – the problem of redistribution

Environmental policy making has always been motivated by the *interests* of the powerful and privileged whether of a particular class or country. In the nineteenth century it was the fear of cholera and other contagious diseases spreading from the slums of the burgeoning industrial cities that led to the various public-health and housing acts and the municipal provision of infrastructure. A venous–arterial system of sewers and piped water emerged in the cities, improving public health but, at the same time, facilitating the spatial social segregation along class lines. In the twentieth century the systematic process of environmental inequality has resulted in locally unwanted land uses (LULUs) becoming concentrated in what have been termed 'peripheral' communities (Blowers and Leroy, 1994). Examples of such communities can be found in most advanced industrial societies, areas with a concentration of high-risk activities such as petrochemicals or nuclear facilities. Among the peripheral 'nuclear oases' are Sellafield in the UK (McSorley, 1990; Wynne *et al.*, 1993), Cap de la Hague near Cherbourg in France (Zonabend, 1989) and Hanford in Washington state, north-west USA (Loeb, 1986). Such communities are the product of a process of 'peripheralisation', reflecting the ability of powerful communities (often acting in combination) to resist LULUs and the powerlessness of peripheral communities to resist them. There is also a pattern of 'environmental racism' whereby environmental hazards such as toxic-waste sites, industrial pollution and pesticide exposure of migrant farmworkers disproportionately affects minority groups, for example in the USA (Cutter, 1993, p. 27).

This process of social and environmental inequality continues; it is merely its scale that has shifted. As de Swaan observes, 'What has lingered on and become acute again are the problems of pollution and environmental protection, but this time at a higher level of integration, that of the national state and, even more pressing, at the supranational level' (1988, p. 142). Heavily polluted areas occur in various parts of the developing world where the need for economic development and dependence on major industries (often multi-national companies) combine to reinforce a condition of powerlessness. The dumping of toxic wastes and hazardous materials in poor countries is an obvious example. In general terms, the dominance and dependence and the power and powerlessness that characterise the process of peripheralisation at sub-national level can also be seen to operate at international level.

Environmental inequalities are an inextricable element of processes of uneven development that exist between communities, classes and nations. With the onset of global environmental changes, it would appear that all are threatened and so there is a mutual interest among rich and poor, North and South in sustainability. But, even with such global threats as climatic change, the impacts are likely to be discriminatory, at least in the short run. The onset of climatic change is likely to be gradual, uneven and to bear most heavily on those countries and communities unable to take defensive actions against drought or sea-level rise. 'For most of the world's people, however, it is not controlling greenhouse gas emissions that is the priority, but ensuring adequate protection from the impact of climate change and sea level rise' (Holmberg *et al.*, 1993, p. 27). Other global processes, notably deforestation, desertification and transfer of hazardous wastes, also have a disproportionate impact on environmental quality and livelihood in the poorer countries. In any case there is a tendency for the advanced countries to give lower priority to those global issues that have least impact on their interests. *If sustainability is to be a politically realistic goal then the inequalities in environment and development must be addressed by the richer nations of the world.*

Of itself, greater equality will not secure environmental sustainability. Quite obviously if all countries enjoyed the wealth of the rich nations the high per-capita consumption of resources and the massive burden of pollution would overwhelm the earth's carrying capacity. Conversely, it is clear that the pressures placed on the resources of the environment in the poorer countries also lead to environmental degradation and destruction. The solution will require totally different approaches to economic development which will be strenuously resisted by those wealthy countries already enjoying a disproportionate share of global environmental resources and imposing a disproportionate burden of pollution and waste on the earth's ecosystems. At the same time, the developing countries are unlikely to accept restraint on their progress so long as the distribution of resources is seen to be blatantly unfair.

Inequality is created by power relationships. In a practical sense it is about control over and access to resources, economic and environmental. It can therefore be addressed by policies of redistribution which transfer resources from rich to poor, by policies of restraint which enforce changes of behaviour on the part of communities, classes or countries or by policies of compensation which seek to mitigate adverse environmental impacts endured by disadvantaged groups, places or countries.

In this section I have examined the fundamental preconditions of policies for sustainable development. They have been considered in terms of the three Es – externalities, evidence and equality – each underlining the need for intervention. Externalities emphasise the need for intervention to secure environmental policies that reflect the public or common interest. The evidence of environmental change is uncertain but this very uncertainty suggests a need for intervention based on a precautionary approach to policy making. Equality focuses attention on uneven development and the need for intervention to ensure redistribution of wealth and power if policies are to be made effective. Now the chapter turns to the nature of the intervention, to the process I have called 'environmental planning'.

ENVIRONMENTAL PLANNING

The goals of environmental planning

All plans are statements of intended futures and, consequently, they must establish their objectives. At the broadest level, the environmental future should be based on the goal of sustainable development. But it is impossible to derive specific policies from so elusive and inchoate a notion. The oft-quoted elaboration in the Brundtland Report, though it clearly expresses the societal aspects of the concept, does not move us towards an operational definition. 'Sustainable development is development that meets the needs of the present without compromising the ability of future generations to meet their own needs' (WCED, 1987, p. 8). This statement begs the question of what is meant by 'development', 'needs' and 'future generations'. Development is a physical process that can, to some extent, be measured but it is also a qualitative, cultural and social concept with values varying over time and space. Sustainable forms of development, therefore, must not only respect the physical limits imposed by the earth's resources but also the cultural expressions of conservation. 'Needs', too, is a relative concept – what are basic needs for one society can be luxuries for another. And the concept of 'future generations' invites us to consider the costs and benefits of present actions for the future.

It is possible to interpret the Brundtland statement in order to give a tighter definition of sustainable development that can inform the purposes of environmental planning. Such a definition, one that identifies the essential physical and social criteria, was adopted by the Town and Country Planning Association (TCPA) in its report, *Planning for a Sustainable Environment*, as follows:

> Sustainable development is development that enhances the natural and built environment in ways that are compatible with
>
> 1. The requirement to conserve the stock of natural assets, wherever possible offsetting any avoidable reduction by a compensating increase so that the total is left undiminished.
> 2. The need to avoid damaging the regenerative capacity of the world's ecosystems.
> 3. The need to achieve greater social equality.
> 4. Avoiding the imposition of added costs or risks on future generations.
>
> (Blowers, 1993a, p. 6)

It should be noted that this definition is anthropogenic, emphasising *human* needs and accepting the possibility of some diminution of natural assets or their substitution. It is, therefore, a 'weak' definition compared to the ecocentric and uncompromising definition sought by advocates of 'strong' sustainability. While recognising the limitations of its pragmatic approach, the TCPA's definition can nevertheless be applied to the problem of environmental planning. Five fundamental goals can be specified. These goals are intended to relate the local to the global, to recognise the social and political determinants of environmental issues and to identify the principles of equality and democratic participation upon which a successful strategy for sustainable development must be based.

The first goal focuses on the conservation of the natural-resource base, setting out some of the basic principles for resource planning.

1. *Resource conservation*: to ensure the supply of natural resources for present and future generations through the efficient use of land, less wasteful use of non-renewable resources, their substitution by renewable resources wherever possible and the maintenance of biological diversity.

The second goal is very much concerned with the land-use planning process itself, focusing on the nature of urban development.

2. *Built development*: to ensure that the development and use of the built environment respects and is in harmony with the natural environment, and the relationship between the two is designed to be one of balance and mutual enhancement.

The third goal deals with the need to prevent or constrain developments and production processes that degrade, pollute or destroy the environment. It establishes the need for environmental planning that focuses on impacts, defines limits and monitors the state of the environment.

3. *Environmental quality*: to prevent or reduce processes that degrade or pollute the environment, to protect the regenerative capacity of eco-

systems, and to prevent developments that are detrimental to human health or that diminish the quality of life.

The fourth goal turns attention to the change in social and economic policy that will be necessary. This would not only be counteracting the ideologies of liberalisation, deregulation, privatisation, individualism and freedom from restraint that have achieved a global dominance, but would require the definition of feasible and politically attractive alternative approaches. Such approaches mean basic changes in patterns of consumption, the allocation of resources and, consequently, of life styles. Environmental planning may be able to contribute to policies for greater social equality but these can only be a part of a much wider change in social values and political priorities. Nevertheless, the goal of social equality is, as was argued earlier, a necessary precondition to securing the co-operation on which sustainable development strategies must be based.

4. *Social equality*: to prevent any development that increases the gap between rich and poor and to encourage development that reduces social inequality.

This leads on to the fifth goal. Fundamental social changes will only be achieved by a combination of comprehension, commitment and consent. People must be fully informed so that they are in a position to accept change and become committed to it. A political as well as social revolution must be envisaged in which democratic participation at all levels is encouraged.

The aim is to achieve the principle of subsidiarity, taking decisions at the lowest level compatible with attaining required objectives. Here, too, the trends seem entirely in the opposite direction. In the UK local democratic participation has been progressively reduced by a process of centralisation, by the dispersal of functions to a raft of unelected quangos, by transfer from the public to the private sector and by heavy restraint on the powers and finances of local authorities. A shift towards sustainable development will involve, as we have seen, a commitment to politically unfashionable notions of collective provision and intervention. Therefore, a fifth goal for sustainable development would be:

5. *Political participation*: to change values, attitudes and behaviour by encouraging increased participation in political decision making and in initiating environmental improvements at all levels from the local community upwards.

Viewed in terms of these five goals, environmental planning becomes a comprehensive social and political process. As a process, environmental planning must be both *integrated* and *strategic*. These are rather unlovely terms but they indicate the essential components of the environmental planning process. The next part of the chapter examines the process, focusing on the UK within the broader global context.

Environmental planning – an integrated process

The environment is integral to human life, health and survival and therefore cannot be separated from other human activities. Environmental processes are integrated in three dimensions, each of which has implications for environmental planning (these are discussed in more detail in the TCPA's report).

The first is the *trans-media* nature of environmental processes. Pollution can adopt different forms and pass through different environmental pathways of air, land and water. For this reason, environmental planning needs to bring together hitherto separate functions of land-use planning whereby local control and environmental regulation covering water and air have traditionally been supervised by quasi-government bodies ultimately accountable to the central government. To an extent the creation of the Environment Agency in April 1996 incorporating the National Rivers Authority, Her Majesty's Inspectorate of Pollution and local authority waste regulation is a step in this direction though it excludes land-use planning.

Second, environmental processes are *trans-sectoral*; in other words they do not respect traditional policy boundaries. It is self-evident that the various sectors in the economy – agriculture, industry, energy, transport, construction – have major environmental impacts, depleting resources and creating pollution, yet the emphasis of policy making is on production rather than conservation. For instance, the decision to discourage out of town superstores was not made until after the vast majority of such stores had already been opened, when it became clear that city-centre commercial interests were being threatened; any cuts in the road programme are likely to arise from financial restraint rather than the need to arrest the growth in traffic or to boost public transport; and the increase in VAT (value-added tax) on domestic fuel in 1994 had far more to do with the budget deficit than energy conservation. Policies for sustainable development must integrate the environment into the vertical economic sectors rather than simply add on an environmental component. At the rhetorical level there is increasingly strong commitment to integration. In the UK tentative steps have been taken towards the so-called 'greening of government' (HMSO, 1991). The government, in its strategy for sustainable development, claims that it 'has long been committed to the integration of environmental concerns into decision-making at all levels' (HMSO, 1994a, p. 197). At the European level, the EC's Fifth Environmental Action Plan, *Towards Sustainability*, emphasises the need for 'integration of environment considerations in the formulation and implementation of economic and sectoral policies, in the decisions of public authorities, in the conduct and development of production processes and in individual behaviour and choice' (CEC, 1992, p. 3).

A third dimension of integration results from the *trans-boundary* nature of environmental processes. The impacts of resource depletion and environmental degradation transgress political boundaries at all levels. At the local level, transport movements and associated pollution cross boundaries, polluted watercourses affect areas downstream, factories spread air pollution and wastes are moved from one area to another. At a regional scale, pollution crosses national frontiers, an invasion that cannot be prevented without international agreement. Globally, trade and investment flows transfer resources and polluting activities from one part of the globe to another. Vertical integration must therefore also occur between different levels of government so that actions taken at one level are compatible with those taken at another.

Environmental plans – a strategic approach

The process of integration between media, between policy sectors and between levels of government indicates a strategic approach to environmental planning. Environmental plans will be needed at every level, each fulfilling three basic functions. First is the elaboration and co-ordination of integrated environmental policies over different time periods and spatial scales. Second is the identification of targets consistent with these policies and methods for implementing them. And third is the monitoring and evaluation of outcomes.

A national environmental plan would translate international agreements and targets into national policies and also provide a framework, targets and means for implementation for subsidiary regional and local plans. Among the targets would be those for carbon emissions, energy consumption, traffic growth, resource conservation, recycling of wastes, air and water quality. Some work has already begun on identifying the types of targets needed. For example, in 1996 the UK government published 120 indicators of environmental trends over the preceding 25 years but only a tenth had targets attached to them. Problems of data availability, compatibility of statistics and the complexity of sustainability make setting targets a difficult task.

> Sustainability indicators need to take account of economic linkages, quality of life and perhaps future welfare aspects, as well as environmental quality ... The challenge is to strike a balance between having a small number so that the main messages are clear, while not oversimplifying the issues or omitting significant areas, or suppressing significant geographical variations.
>
> (HMSO, 1994a, p. 220)

Such variations can be dealt with through regional and local environmental plans. In the case of the UK the regional tier of government is a missing dimension. Yet it is at this level that the links between the spatial economy and the environment can be tackled strategically to articulate the relationships and balance between the economy and the environment. It is here that sectoral policies for agricultural development, energy, resource conservation, the physical and social infrastructures of housing, transportation, industrial development and waste management can be brought together and related to environmental policies aimed at reducing waste and pollution and protecting natural resources. The present *ad hoc* advisory system of regional planning is inadequate for the task. The myriad of quangos responsible for the delivery of a wide range of policies at the regional level are not accountable, nor are they democratic. At the very least, these regional functions need to be effectively co-ordinated and focused clearly on goals and targets of sustainable development. But, in line with the goal of political participation, a regional elected tier of government will be necessary to ensure that national sustainable development policies are effectively integrated across the range of sectors at the regional level. The regional environmental plan would, in its turn, provide the framework of broad policies and targets for the subsidiary local environmental plans.

These local plans, developed by the local authorities, would have two functions. One is *responsiveness*, to ensure the effective implementation of national and local policies for sustainable development. But local government is not simply 'a policy vacuum waiting to be filled by initiatives from the centre' (Ward, 1993, p. 469). Its other function is to stimulate innovation and local initiatives that contribute towards the goals of sustainability. 'As the level of governance closest to the people, local authorities play a vital role in educating, mobilizing and responding to the public to promote sustainable development' (UN, 1992, p. 233).

There is, of course, much evidence of vigorous activity at the local level with the publication of environmental charters, action plans and environmental audits. Some local authorities have achieved a reputation for innovation; Sutton, for example for its early (1986) adoption of an environmental strategy, Lancashire for its influential environmental action plan and Leicester, Middlesbrough, Leeds and Peterborough for developing the networks and activities that have designated them 'Environment Cities'. The majority of local authorities are engaged in environmental planning of one form or another (Raemakers, 1992) and all are expected to define their local Agenda 21 by the end of 1996.

Sustainable development has become the *leitmotif* of development plans. The Dutch national environmental policy plan is the archetype of a national plan with its analysis of problems, identification of issues, integration of different scales and development of objectives, policies and indicators within specific time-scales. It emphasises the interaction of natural, social and economic functions of the environment, arguing that the 'main objective of environmental management is to preserve the environment's carrying capacity for the sake of sustainable development' (NEPP, 1989, p. 15).

At a more local level, Bedfordshire's structure plan represents the new thinking that goes beyond mere rhetoric and defines goals for a range of policies with specific targets to be met. Examples of these targets for the period up to 2011 are for 80 per cent of new built development to be within urban areas and main transport corridors; loss of greenfield land to be reduced by 50 per cent compared to 1986–91; a reduction of 50 per cent in the amount of derelict land and vacant buildings; a 100 per cent increase in the level of energy produced from non-fossil fuel sources; a 25 per cent reduction in the amount of waste disposed to landfill; a reduction of CO_2 emissions from buildings, industry and transport to the 1991 level by 2001 and a 15 per cent reduction thereafter; a target of 50 per cent of all journeys to work within urban areas by public transport, walking and cycling; and a doubling of woodland in Bedfordshire by 2015 (Bedfordshire County Council, 1995). Bedfordshire's plan is a good example of local government seizing the initiative and attempting to set the agenda for sustainable development.

As the Bedfordshire plan recognises, local councils can only affect a few or part of these targets. Indeed, the loss of local-government powers and the fragmentation of strategic local planning that is likely to follow recent local-government reform have seriously weakened the influence of local government in the UK. As the 'democratic deficit' grows so the scope for turning local initiative and commitment into effective action diminishes. In any case, sustainable development policies cannot be achieved by government alone, at whatever level. It can only be realised if all responsible bodies and the private sector act in conjunction. The final section of this chapter sets out the political conditions that need to be met to achieve sustainable development.

THE CONDITIONS FOR CHANGE

The role of the nation-state

The nation-state is the only form of political organisation that occurs throughout the world. Ultimately the nation-state is the authority through which policies for sustainable development will succeed or fail. And yet nation-states are not all-powerful and there are vast differences among them. Although nation-states claim sovereign authority over defined territories they are very unequal in terms of power, competence and legitimacy. Some states are riven by internal conflict, indebted to donor countries and lack the resources, commitment and ability to ensure effective implementation of policies. The environment can become progressively degraded and its resources plundered without regard to any criteria of sustainability. Other, more powerful states, are able to exert their authority without regard to external pressures and can legitimate policies or processes that may prove environmentally destructive within their borders or that impact on neighbouring countries and the global environment. For example, the Scottish

Flow Country, a wilderness area of international importance, can be destroyed or the Yangtse Gorges flooded without external interference. The trans-boundary transfer of pollution amounts to an invasion by one country on others against which there is no conventional defence.

In practice, nation-states, even the most powerful, do not have supreme authority within their boundaries. There have always been tensions between interdependence and sovereignty. As Potter observes, 'States are not sovereign, but they are powerful, and in some settings very powerful, within international political contexts' (1995, p. 107). Environmental political and economic processes transcend state frontiers, linking the local to the global. States are therefore constrained not only by the political processes within their borders but also by transnational political and economic processes. States acting individually or together are able to influence those processes. In some cases, governments acting in concert can influence environmental policies as did, for example, the group of African countries that sought, through the Bamako Convention, to ban the trade in hazardous wastes that had become an environmental threat to them.

Intergovernmental organisations (IGOs) have an impact on the environment. Some of these deal directly with the environment; for instance, the various conventions, protocols and other agreements dealing with regional and global environmental problems culminating in the Rio conference. Others, concerned with development (World Bank, International Monetary Fund, multilateral development banks, Food and Agriculture Organisation) or trade (World Trade Organisation) exercise great influence and power within some of the developing countries and have been criticised for their adverse impact on the environment (Bramble and Porter, 1992; George, 1988; Lang and Hines, 1993).

Non-governmental organisations (NGOs), too, exert considerable influence within countries. Environmental NGOs act as lobbies within the developed countries and within the developing world, and NGOs (both indigenous and those based in the North) are often involved in the implementation of programmes of environment and development (for example, they are recognised as important agencies for implementation of the Convention on Desertification). Some NGOs operate on both the national and international level and, in the environmental field, Greenpeace and Friends of the Earth are the most prominent examples. But there are also coalitions of NGOs operating in networks to bring influence to bear both on individual countries and on the international environmental agenda.

Multi-national corporations (MNCs) are another major source of influence over economic and environmental policies; they exert a global reach over the exploitation of resources, the distribution of investment and the scale and location of environmental degradation. Some poorer countries are dominated by the operations of the MNCs and, even in the rich countries, they have a major influence on governments. Though bound by environmental regulations, MNCs will bring pressure to bear on governments and the international community to limit the impact on their freedom of operation. They advocate deregulation, liberalisation of trade and the operation of an unrestrained market – all of which tend to undermine environmental protection and resource conservation.

These various transnational political entities – IGOs, NGOs and MNCs – pull in different directions, exposing the fundamental conflict in the contemporary world between environment and development. The nation-state remains the key to integrating environment and development in the interest of sustainability. This involves surrendering some power upwards to ensure international agreement on sustainable policies and devolving some power downwards to stimulate participation and innovation as close to the communities affected as possible. In a world where the capacity of nation-states varies so widely, it may prove difficult to achieve any consistency over policy making let alone ensure consistent and continuing implementation. Added to this is the problem of shifting the emphasis of economic management from liberalisation to greater intervention to protect the environment.

Equality and compensation

As was argued earlier, sustainable development is only going to be politically possible if there is greater social equality both within and between countries. This implies some transfer of resources, technology or political participation from the rich areas to those communities and countries experiencing environmental disadvantage.

There are three principles under which compensation will be necessary to achieve greater environmental equality. First, those communities and countries ravaged by exploitation of natural resources or degraded by pollution should be compensated for the disproportionate burden of risk that they bear. As locating LULUs becomes more difficult, it will become not only morally right but politically necessary to compensate such communities in the form of tax relief, economic regeneration, community projects, provision of mitigation measures and greater political participation.

Second, there will also need to be compensation for communities or countries where development is restricted for reasons of sustainability and where economic hardship results. It is not necessary to compensate, say, landowners, developers or farmers who are prevented from opportunistic changes to a more profitable use. There is a distinction to be made between denial for public good and the opportunity for private profit. But it is necessary to compensate those areas or countries that would otherwise be unable to avoid the depletion of resources or destruction of the environment. This amounts to a transfer of resources to those areas that hold resources necessary for the survival of ecosytems. An interesting counter-principle also applies. If some parts of the world must be held in reserve as common resources for sustainable development, then those areas that are developed should also be used for the common benefit. To put it plainly, 'The conservation effort of the world should occur mostly in undeveloped areas. But this is not to let people in developed nations off lightly. Their burden is providing a place for people to live' (Luper-Foy, 1992, p. 62).

Third, compensation should, in principle, also be provided for those countries or communities that suffer from environmental damage inflicted on them by neighbours or more distant polluters. Such externalities should be compensated by invoking the principle of the polluter pays. In practice this may prove difficult to determine, as we have seen. In many cases it is hard to establish responsibility for cause and to calculate the costs of damage. Polluters will naturally wish to evade or to minimise their culpability. In other cases, polluters may simply lack the resources to provide compensation. At the local scale, enforcing the polluter-pays principle might risk running a company out of business. At the global level, even individual countries might find it impossible to compensate for the export of their pollution elsewhere. The most dramatic instance of this so far has been the case of Chernobyl where the Soviet state and its successors were quite unable to pay for the damage caused to other countries and needed technical and financial assistance to improve the safety of their nuclear power plants. Global environmental risks must be dealt with by international emergency planning and funding both for preventative and remedial action.

Grounds for optimism?

The political conditions for achieving planning for sustainable development outlined here may seem hopelessly unrealistic. It seems far more likely that the relentless process of market liberalisation, dominance of multi-nationals and the pursuit of national self-interest will accelerate the depletion of resources and the degeneration of the environment. There would appear to be little hope of compensatory shifts from rich to poor in order to mitigate or avoid further damage. Any restrictions on the commercial freedom enjoyed by the wealthy parts of the world will be strongly resisted. Policies for sustainability are likely to be overwhelmed by the onward surge of competitive, exploitative and unsustainable development.

Yet it would be wrong to dismiss planning for sustainable development as an unattainable ideal. There are trends already under way that may make it increasingly practical. Already, as we have seen, the political revolution of the late 1980s, largely unanticipated, swept away an empire and initiated a reappraisal of international expectations. It is still far too early to comprehend the consequences, but the changes at least raise the possibility of greater co-operation on global environmental issues.

At the same time we are experiencing a profound change in information technology as far reaching in its implications perhaps as the introduction of printing over 500 years ago. Within affluent societies computers have already become an essential tool of management and education. With the advent of the information superhighway the social transformation already under way will vastly accelerate. It will become possible for anyone linked into the network to have instant access to a diversity of information from around the world. The ability to receive, interpret and transmit information, knowledge and decisions will transform working practices and shopping habits. The changes in travel habits as commuting and shopping journeys reduce and leisure trips increase has obvious consequences for spatial patterns of development.

The implications for sustainable development of the information revolution, however, go far wider than simply shifting the volumes and distances of travel. The wider availability of environmental information could help the environmental planning process. But, most importantly, information will become more accessible and therefore less centralised and hierarchical. The power of holding and witholding information will be undermined and, with it, centralised bureaucratic structures and secretive decision-making structures will succumb to more democratic and diffused networks. The basis of the nation-state's power will be challenged as information transcends frontiers. Information networks with truly global reach could encourage co-operation and mitigate the competitive conflict that has characterised the rise of the nation-state. Already the linkages achieved by scientists and NGOs have demonstrated the potential power achieved through the dissemination of environmental information. Such a shift in the basis of power would be truly revolutionary. It would certainly facilitate the kind of political participation that, it was argued earlier, is a precondition for environmental planning.

All this may seem a long way off at present. As with all revolutions the immediate beneficiaries are likely to be the affluent and powerful. Patterns of social inequality which flourish under existing social arrangements are likely to persist so long as there is unequal access to information. But at least there is the potential for widespread, perhaps ultimately universal, access. When that occurs, the need for environmental planning for sustainable development will not only be recognised but a transformation in social relations will have occurred that will make it possible.

4

PLANNING, PROFESSIONALISM AND SUSTAINABILITY

Bob Evans and Yvonne Rydin

During the last five years there has been a sea change within the policy area of land-use planning. The post-Rio 'new environmental agenda' of sustainability[1] has permeated most corners of the planning profession and virtually every planning policy document now emerging from central and local government makes a ritual nod in the direction of the Brundtland definition of sustainable development. Moreover, British planners have, in general, welcomed this new environmentalism. Indeed, both the profession and central government have seen the land-use planning system as having a special role in achieving sustainability, charging planners with a major responsibility for this overall goal.

However, this changing policy agenda raises some important questions concerning the role of the town-planning profession. To what extent does this occupational group possess the skills and expertise necessary to implement this new environmental agenda? Furthermore, can the perspectives contained in the work of Healey, Chambers and others provide an adequate model for professional planning practice as we move into the millennium? This chapter seeks to address these issues through an examination of the UK planning profession and of the expertise and skills that might be appropriate for the delivery of the Rio sustainability agenda.

Before we outline the bones of our argument, we need, first, to establish our starting position. The term 'planner' tends to have a general and non-specific usage, and we should emphasise that we use the term here to apply to those people who are professional planners, that is, in Britain, members of the Royal Town Planning Institute. Clearly there are many people who are not Institute members, who are employed as 'planners', environmental or otherwise. However, as will become apparent, our focus in this chapter is the professionalisation of knowledge and the consequences of this for environmental policy making and implementation.

We accept the argument made elsewhere that the skills and expertise that underpinned the professionalisation of planning in Britain during the first half of this century are now increasingly irrelevant (Reade 1987; Evans 1993). As has been argued in the first chapter of this book, 'classical town planning', the process of town design, and the expertise associated with it, has little purchase on contemporary environmental and land-use problems.

Furthermore, we broadly accept that the contemporary profession and practice of British town planning is experiencing a crisis of legitimacy for a variety of reasons including:

1. the lack of any coherent theoretical underpinning, which has increasingly undermined the claim to technical and professional authority;
2. the loss of an interventionist capacity during the last two decades, which has seriously impaired the planning system's ability to counter market forces through 'positive planning';
3. the claims that the land-use planning system operates in 'the public interest', which need to be set against the widespread perception that planning inevitably tends to benefit land and property owners and the educated and articulate middle class at the expense of other members of society.

Those within the planning profession, of course, bemoan the marginalisation of planning into a quasi-legal process of regulation and limited negotiation with the market. More controversially, we argue that it may be more appropriate to conceive of the planning profession as playing a circumscribed, though important, part within a broader environmental policy process. Furthermore, while the professionalisation of planning offers certain opportunities for raising the profile of the new environmental agenda, the professionalisation project can also serve to hamper moves to meet the challenge of environmental sustainability.

Given these starting points, the chapter proceeds in the following way. We first consider the case that planning is a form of general competence that can be applied to sustainability issues as readily as any others. This involves a brief overview of the history of normative perspectives in planning and an account of the specific competences of land-use planning. We then turn to the argument that planning is the repository of a specific set of values, a 'welfare profession'. This involves a review of the arguments on the nature of professionalisation. Our third and final section considers the expertise that is demanded by a sustainability agenda and makes comment on the role that professional planners can realistically play. This involves a consideration of the political as against the technical role of planners.

PLANNING AS A GENERAL COMPETENCE

We have said that our focus is on professional planners – urban, town or land-use planners – but such planners have tended to define themselves as holding more general skills, expertise and competences. This is the result of an historical broadening of the role of planners as the profession has redefined its expertise over time in response to social, political and economic pressures (Hague 1984; Reade 1987). It is possible to characterise these changes as follows.

Planning as reform and town design

This is the legacy of the 'founding fathers' (*sic*) such as Howard, Geddes, Sharp and Abercrombie. Overwhelmingly physically deterministic, this earliest planning blended social concern with an architectural appreciation to produce the 'art and science of town planning' – a combination of artistic inspiration, grand narrative and social reform which was to characterise the planning profession for well over half a century. The planners' expertise was alleged to be 'town design' – the capacity to create an environment that was functionally appropriate and aesthetically pleasing – although the knowledge and theoretical foundations upon which the skills necessary for this might be based were not clearly specified.

Planning as technical rationality

As statutory land-use planning expanded in Britain during the 1950s, 1960s and early 1970s, town planners increasingly became involved with massive urban redevelopment and slum-clearance schemes of great complexity. Strategic planning at the regional and sub-regional level also became a high-profile planning activity. Associated with this was a change in emphasis in terms of the expertise claimed by the profession. Planning began to claim a managerial competence – the ability to oversee a team of professionals in order to co-ordinate and implement policy – and a strategic competence – the ability to develop plans, policies and strategies for the future to meet specified goals. This ambitious remit was reflected in the focus upon systems theory which, it was argued, would enable planners to harness the potential of computing in order to model reality (McLoughlin 1969).

The development of systems theory in planning seemed to offer the prospect of a planning based on specific technical skills in information technology applied to, but not limited only to, land use. Planning theory developed generic models of decision making, principally procedural planning theory (PPT) or the rational comprehensive planning method (RCPM). However, it soon became apparent that this approach was impossibly ambitious and naive in its attempt to technicise complex social and political processes (Healey *et al.* 1982).

Planning as negotiated, regulative order

During the last two decades, planning has moved away from these grandiose claims and has moved towards a lower-key claim to an expertise in managing and manipulating the statutory planning system for both the public and private sectors. Here the planners' role is to oversee the legislative process and to mediate between competing interests. The claim to expertise here is based upon a knowledge of the policy process in managerial and political terms, and of procedures and case law, linked to knowledge of the economic processes by which urban development is generated and shaped, and a capacity to mediate.

This latest phase of change is a response to the criticisms of planning as technical rationality at both a theoretical and practical level. It is also, of course, a response to the 'space' left to planning by the restructuring of the Thatcher years. However, while it appears that planning now occupies a more specific space, this formulation still sees planning expertise in general terms, as skills that may be found in any policy area. This normative view allows planning to move into other policy areas, such as the 'new environmental agenda'. This seems destined to form the next phase.

Planning as environmental management

This model emphasises the role of the planner as a manager, bringing the environmental-policy process together, and as such it is a perspective that has its roots in the 1960s and 1970s, perceiving the planner as the leader of a team of other professionals with a strategic 'eye', though refined by the new negotiative and mediation skills emphasised during the 1980s and 1990s. The contemporary variant thus has its roots in both the technical rationality and the negotiated regulative-order interpretations of planning expertise. The planning profession is seen as having skills of a quasi-technical nature in fields such as environmental assessment or the construction of sustainability indicators, whilst also contributing special skills in mediation between state agencies, citizens and other players in the environmental arena. The expertise claimed is thus partly technical and partly managerial.

The 'new environmental agenda' has provided an opportunity for professional planners to claim a competence in environmental management, partly through the development of the Local Agenda 21 (LA21) process, and partly as a result of the 'greening' of land-use planning as exemplified in recent Planning Policy Guidance Notes. However, while there is undoubtedly a 'policy window' for professional planners here, there is a question mark over their abilities to continue to take on such a broad role. We comment on this further below in the context of how contemporary normative planning theory fits with the sustainability agenda. For the moment we would note that, drawing back from an ever-broadening role for planning, there are two specific areas of planning competence which seem to us to have maintained a thread of continuity throughout these phases and which seem to provide the bedrock of planning expertise, from which claims to broader competence should be judged.

These two areas of competence are knowledge of the administrative systems of planning and a concern with urban design. The first of these probably reflects the views held by many practising planners in both the public and private sectors. Planning here is mainly a quasi-judicial process of implementing the town and country planning legislation and associated governmental guidance through plan preparation, other policy documentation and development control. Expertise is defined as knowledge of the procedures, legislation, precedent and 'case law', with experience and 'on-the-job' learning accorded high status. This planning work has both a public- and a private-sector dimension. In the latter situation, planners may seek to utilise a knowledge of 'the system' in order to secure consents for clients. In contrast, public-sector planners may utilise somewhat similar knowledge and expertise to defend 'the public interest' against the market. This somewhat gladiatorial depiction characterises many public inquiries and other planning occasions, though moderated at times by the pursuit of negotiated outcomes.

The second area of competence is more problematic. The architectural and design legacy of what we have termed 'classical town planning' may be found in the urban-design elements of contemporary planning. This type of work focuses upon the creation of 'spaces and places' through either the manipulation of planning controls, as for example in conservation areas, or through the design of new or reshaped localities. The skills here are seen as an amalgam of the traditional 'art and science of town planning' – the knowledge of how 'to plan' – with certain architectural skills. As with architecture, the artistic attributes of flair, imagination and originality are seen to combine with a sense of scale, humanity and utility to secure liveable and aesthetically pleasing environments. Clearly many architects would also claim to possess these capacities and skills; it is unlikely that many professional town planners not also trained in architecture have them to the same degree. Nevertheless, re-interpreted at the site rather than the urban level, this area of planning work has considerable importance in that it describes the job content of much development control and negotiation on development projects, and it also relates closely to lay conceptions of 'what planners do'.

These two core competences of administrative regulation and urban design are, however, a long way from the broader claims to a general and generic planning competence. We would argue that the narrower definition should hold sway, at least until the more general claims can be assessed in the context of the sustainability agenda. We now turn to another key claim of the planning profession: to operate in the public interest, in support of liberal values – to be a welfare profession.

PLANNING AS A WELFARE PROFESSION

British planning has historically been associated with a reforming idealism and professed altruism, since its nineteenth-century roots in public-health concerns. As a result, the literature concerned with land-use planning regularly implies that 'planning' has deeply entrenched characteristics that distinguish it from other areas of public-policy making. As far as we can see, there is no justification for this position. Land-use planning is clearly an important and essential area of public policy that no modern society can afford to ignore. However, it is simply *that* – one area of public policy amongst many, and there are no grounds for representing it as different or special, inevitably imbued with particular ethical, moral or reformist characteristics.

This also calls into question the equally frequently asserted characteristic of planning – that it is a progressive activity conducted in the 'public interest'. This conflation of the *process* and the *ends* of policy is a confusion which, as Reade (1987) argues, has served the profession well, but the available evidence clearly shows that land-use planning tends inevitably to benefit the articulate and educated property-owning middle class, rather than serving some variably defined public interest (Simmie 1981, 1990). Planning operates in areas of conflict over land use and the very definition of the public interest in such circumstances is highly contested.

Following from this, we suggest that, contrary to claims regularly made by the planning profession, there is no such thing as 'good planning' in any objective sense, since the theoretical criteria upon which such a judgement might be based hardly exist. Whilst 'good engineering' or 'good dentistry' might be identifiable, 'good' or 'bad' planning are conceptions determined mainly by social and political circumstances and personal preferences, and not by objective and 'scientific' criteria.

A road bridge or a dental filling may fail: planning failures are socially identified and highlighted. This is equally true in relation to the sustainability agenda, given the social construction of many of the key terms within the agenda and the uncertainties that exist in relation to many of the scientific claims. We return to this in terms of considering the technical and political sides of planning for sustainability in the final section.

The underlying point here is that it is important to recognise that professions do not exist in a social, economic and political vacuum. Professions are a form of occupational control – in Johnson's famous phrase, a 'profession is not, then, an occupation, but a means of controlling an occupation' (Johnson 1972: 45). Crompton reinforces this point when she argues that professions are but one mode of what she terms 'moral regulation of expert labour' (Crompton 1990). They exist to protect the vulnerable non-expert from the expert and to regulate the occupation in the interests of the state, the client, the occupation group itself or some combination of these three. Larson argues that the process of professionalisation must be understood as the process through which the producers of special services seek to constitute and control a market for their expertise (Larson 1977). Not all groups achieve this privileged status, and many aspiring groups fail. The key factor in this success or failure is state support, in that professional status is granted either explicitly or implicitly by the state.

Professionalisation centres on the claims of an occupational group to exercise control over a particular area of knowledge or expertise. Experts are people with a claimed expertise in an area of 'legitimate knowledge', that is, an area of knowledge that is approved, sanctioned or sponsored in some way by the state. As Wilding points out:

> On its own, expertise does not bring power – it has to be useful expertise which is valued by government, and it has to provide guarantees that it will be used only for achieving acceptable purposes in acceptable ways.
>
> (Wilding 1982: 75)

Processes of the codification and credentialisation of knowledge into a recognised 'cognitive competence' thus underpin the existence of a professionalised occupational group of experts. Such knowledge may be created, codified and made exclusive. Equally, as Abercrombie and Urry point out, knowledge may be appropriated from the direct producers and utilised by other groups for their own purposes (Abercrombie and Urry 1983: 92).

Thus the main area of professional advancement 'is the capacity to claim esoteric and identifiable skills – that is to *create and control* a cognitive and technical base' (Larson 1977: 180, emphasis added). Yet, as we have seen, beyond a knowledge of the operation of the procedures and the legislation of town and country planning and a degree of design competence, the specific cognitive competence and expertise of professional town planners is extremely difficult to specify. Surprisingly perhaps, planners even seem unable to do this themselves (Evans 1995b). Hence the professionalisation of planners is inherently problematic.

But professionalisation does not rely *only* on claims to expertise. Professions are also characterised by their assumption that they are undertaking activities of social worth. Summarising the prevailing sociological literature, Freidson argues that 'knowledge and skill cannot advance the *necessary and desirable ends* of sustaining and enriching life without being institutionalised in some fashion' (1984: 25, emphasis added). Drawing on extensive historical research, Larson further notes that 'Professional reformers logically defined expertise as that which they did or thought worth doing' (1984: 34). Indeed, earlier studies of professions were limited to identifying the 'traits' of professionals such as altruism and social responsibility and classifying the scientific nature of their knowledge base (Macdonald 1995).

Therefore the statement of desirable values and the professionalisation of claims to expertise are wrapped up together. This is not to suggest that professions, as organisations or individuals, should not make statements in support of welfare reformism. It is our view that as many people as possible should make such statements! And indeed, because of its historic position, the planning profession provides a cultural space in which such statements can be heard. But these pronouncements cannot be taken as disinterested; they reflect and contribute to the profession's self-definition and there will inevitably be a tension between statements of values and the social positioning of the professional. This relates as much to pronouncements by the planning profession in support of sustainable development or sustainability. This support is part of an attempt to revive the campaigning spirit of a hundred years ago but it is also an attempt to reposition the planning profession centrally within the public and private sectors, to enlarge their claims to expertise and to maintain a broad view of their competence. Our view is that such an attempt needs to be measured against the sustainability agenda very carefully. The third and final section attempts this task.

PLANNING AND THE NEW ENVIRONMENTAL AGENDA

In this section we identify three areas where the new environmental agenda requires expertise and skills and we consider the role of the planning profession in relation to them. In doing so we consider two contemporary normative models of planning and professionalism, as developed by Healey and Chambers, and raise issues of the technical/political interface in planning practice.

The demands of a holistic approach

The first area where the goal of sustainability makes new demands is in terms of requiring an integrated, holistic approach that goes across accepted sectoral and organisational boundaries. The holistic basis of ecological thought is well established (Dobson 1995b); in the UK, the managerial equivalent in terms of the policy process has been restated from the Brundtland Report onwards by, for example, the call for environmental planning as discussed by Blowers in Chapter 3 of this book. Yet it remains very difficult to break down organisational and occupational barriers to integration. Every unit would rather claim the holistic approach than contribute only a part to the whole. This applies as much to land-use planners as to other occupations, departments or policy areas. Land-use planning clearly has a role to play *within* an integrated environmental policy; it cannot stand for that policy.

To fulfil this role, professional planners need to be redirected towards a new set of goals with environmental sustainability alongside social and economic goals, possibly even in a pre-eminent position. Planners will also require new knowledge for this role. In part such knowledge is procedural, adding to the core competence of administrative regulation. In part it involves knowledge of environmental impacts and ways of conceptualising them, on which strategic and project-based environmental assessment and auditing can be undertaken. This is a new demand on planning education, not so much teaching planners how to predict these impacts but enabling them to know when and where to obtain advice on the nature of such impacts.

There is also a need for policy evaluation to ensure that a holistic approach is being taken. But this should clearly operate at a higher level than any planning department or equivalent; it needs to be an overview. Interestingly, anecdotal evidence suggests that professional planners may be rising within organisations to take on such a role. In this case the skills learned at planning schools may continue to be useful but in general we are dealing with abilities broader than land-use planning, largely acquired post-qualification.

The demands of environmental science

As planning has moved away from a focus upon questions of 'amenity' towards a focus upon environmental sustainability, scientific argument has achieved increasing prominence (Myerson and Rydin 1996). The centrality of scientific expertise to identifying the problem and suggesting policy options can hardly be overstated. Planners clearly do not possess such environmental scientific competence and cannot be expected to obtain it. At most they can hope to understand the lay communications of technical reports produced by environmental experts in scientific specialisms. However, this could be stating the planner's role (or indeed the role of any non-scientist) too baldly. The sociology of science literature warns against taking the conclusions of the scientific community as given (Wynne 1994). It reminds us that science is also organised along professional lines and that scientific knowledge, while undoubtedly about a material world, is socially constructed and communicated. There are also inevitable uncertainties in the scientific 'knowledge' that is offered to us, uncertainties over ranges of global warming, the significance of biodiversity, the accuracy of impact measurements and so on.

In this context, claims have been made for a new form of knowledge production. Gibbons *et al.* (1994) argue that the nature of knowledge production is shifting from one mode to another and this favours a shift towards expertise generated in transdisciplinary contexts, problem-solving oriented, socially accountable and transient. Thus focusing on the new mode of knowledge production, Gibbons *et al.* claim that Geographical Information Systems (GIS) and modelling environmental data 'have literally changed the way of seeing and practising regional planning' and have opened up new channels of communication and transdisciplinary research (1994: 39). It would seem that land-use planning, with its spatial focus, could contribute in a transdisciplinary context to environmental scientific knowledge itself.

The demands of the political process

The activities surrounding Local Agenda 21 in Britain have given a fresh focus to the environmental agenda, one which goes beyond an intra-governmental attention to a new policy issue. LA21 has restated the concern with sustainability as the remit of a wide range of actors within the local community/ies. Thus environmental policy becomes an inherently and explicitly political activity involved with different groups within society, addressing their needs and concerns, and relating these to the sustainability goal. The role of the professional planner is now focused outwards and two bodies of theoretical work suggest themselves as helpful in reconsidering this new role: the communicative/collaborative planning theory of Patsy Healey (1993, 1997) and the work on new professionalism by Chambers (1986, 1993).

Healey has drawn on the more developed planning theory and practice literature from North America, itself built on earlier experiences of advocacy planning and environmental mediation. Grounded in the critical theory of Habermas and, more recently, argumentation theory (Fischer and Forester 1993), this seeks to find a role for planning in negotiating a communicative rationality, in shaping forums and spaces for different voices to be heard, in shaping strategies to speak to different groups and actors, and in using communication to build alliances. Thus the negotiated regulative order acquires a postmodern concern with polyphony (Mazza and Rydin 1996).

This normative model of planning practice is attractive, but the shift from negotiation and regulation to communication implies new claims by planners. In the former approach, planners' expertise concerns organisational manipulation and negotiation. The skills involve interpreting rules and regulations, not giving explanations, to follow Svensson's distinction (1990: 50). The expertise is organisationally bounded and founded on how to lead or delegate, how to take policy decisions, how to employ and start projects and how to change organisations. But when one moves to the communicative model the transformation of expertise is marked. Planners are here facilitators, expert in argumentation, the use of language and persuasion, and sensitive to the needs of a range of groups in society. While planning schools would hope to sensitise their students to the perceptions and needs of different social groups, the positive abilities of advocacy and facilitation are rarely dealt with in any depth and indeed conflict with some of the more positivist and technical aspects of the training.

Allied to this issue of the shifting expertise base is the changing nature of the relationship between planning and politics. In a purely administrative regulation model it was clear that professional practice was distinct from political activity. The public and other groups in civil society were consulted for a variety of reasons. Consultation set and refined the policy objectives, objectives which were properly determined in the political arena and which professionals sought to implement (a view effectively debunked by Barrett and Fudge (1981)). But consultation also brought valuable information into the policy process and eased implementation of the strategy devised by planners by providing advance warning, modifying the details of the strategy and legitimating it (not necessarily all at the same time).

With the shift to mediation, planning found itself entering the murky waters of direct political involvement. Mediation is after all a political task. Planners were here engaging with civil society, with vested-interest groups and seeking to find a path between them. That path was presumed to be chosen on the basis of balance (a pluralist notion of governmental activity) or some pre-set policy objectives as in the previous model. The chances of achieving either in a situation where planning and politics overlapped were not high. And this problem is compounded in the communicative model. Here planning has become effectively dissolved into the political arena. Environmental planning becomes explicitly and entirely a political process of talking, hearing and arguing. Planning is not about decision making but evolving a consensus. Planners are in amongst the other groups in civil society in this scenario.

Even the advocates of this model of planning find difficulty in fully accepting all its implications. In a statement for planning as argumentation, Healey refers to development plans as expressing assumptions about various social and spatial relations, rather than knowledge, and yet refers to the preparation and use of a plan as being 'much more than [but not distinct from] a technical bureaucratic exercise' (1995: 255). She recognises that a plan based on open democratic argumentation may involve many voices and yet says that 'This diversity tends to undermine the authority of the plan'; why does a communicative plan need authority? For Healey, development plans are 'used to express and take control of the agendas with respect to the management of environmental change in localities by different groups' (1995: 256); here, planning is not just communicative but it also clearly aimed at achieving some material change. Communicative planning for sustainability would encounter these same tensions.

So we turn to Chambers's work for a different, less disengaged perspective. Chambers starts from the recognition that a recurring theme has been that 'welfare' professionals, often acting with the best of good intentions, have frequently contributed to the immiseration of the very people they have been trying to help. This might be due to ignorance or insensitivity, or, as famously argued in the case of planning by Davies, to the idealistic naivety of 'evangelistic bureaucrats' (Davies 1972). Professions have also been criticised for 'disabling' non-professionals through their control of knowledge and expertise (Illich 1977). Healey's work can be seen as an attempt to counter such criticisms from within the planning profession. Chambers, writing in the context of rural sustainable-development policies for the South, argues for a new paradigm for professionalism. He notes that what he terms 'normal professionalism' has been responsible for the failure of many, if not most, attempts to secure effective economic and social development in the 'underdeveloped' South (Chambers 1986, 1993). Chambers sees normal professionalism as representing a set of knowledges, values and power relationships that conspire to deliver inappropriate and ineffective short-term policy 'solutions'.

In contrast, he argues for a 'new professionalism' which he sees as reversing the roles and power structures of normal professionalism. New professionalism would not simply import the 'core values' of the developed centre, but would seek to value and learn from the knowledge of poor people, and to encourage local initiative and local empowerment. Broadly, his position accords with the notions of empowerment and capacity building that are central to the philosophy of the Rio Agenda 21. Chambers's argument is synonymous with what might be termed the 'alternative professionalism' position in planning – the belief that planners and planning should serve 'the people'. This position has been a recurrent subsidiary theme in planning during the last two or three decades, ranging from the call to planners to be 'bureaucratic guerrillas' (Community Action 1972) to the less extreme agenda of the Royal Town Planning Institute's Radical Institute Group of the late 1970s. Clearly, in this perspective, political commitment is as important as any technical knowledge or expertise.

This raises a new set of questions for such a restructured planning practice. As we have emphasised, professions are state sponsored or approved and, following Johnson (1972), Larson (1977), Crompton (1990) and so on, it is clear, as has been argued above, that they exist to regulate expert labour in the interests of sections of the public, the state, the profession itself or some combination of these. Given this, it would be naive to imagine that it is possible to restructure or reconstitute the character of a profession, and in particular a 'welfare profession' such as planning, without threatening its very existence. As both Larson and Johnson have pointed out, occupational groups can easily lose the advantages of professional status, and this is most likely to occur if the profession is no longer perceived by the state as useful or legitimate. This is the problem with the new professionalism position since adopting a campaigning or confrontational stance would invite censure.

CONCLUDING ON DEPROFESSIONALISING PLANNING

Despite the apparently critical tone of the preceding pages, we wish to argue that many of the core skills and capacities held by planners are essential for dealing with this new environmental agenda. However, we also wish to suggest that changes will be required both in the organisation of expert labour and in the educational structures and programmes that prepare individuals for work in this field.

It must be recognised that professionalisation necessarily implies depoliticisation. Larson argued in the 1980s in a US context that 'the same deep structures underlay the expanding role of experts and the drastic impoverishment of political life and vision in advanced capitalist societies' (1984: 30) and that '[t]he inevitable recourse to scientific and technical expertise is one more factor that reduces legitimate citizen participation in decision making' (1984: 39).

By extension, the shift from the traditional models of professional planning to the communicative model or a new professionalism is likely to be associated with a deprofessionalisation of planning. This arises from the change in the nature of the expertise claimed and the changed relationship with civil society outlined above. Looking at the changes of the 1980s, Johnson argues conversely that the Thatcherite project of deprofessionalisation has politicised environmental and spatial issues by removing them from the arena of 'neutral' professional judgement (1993: 139). This is not necessarily, by any means, a problematic change.

It is also important to recognise that many current views of planning confuse the process and goals of planning and this is particularly the case with the adoption of sustainable development. Linking sustainable development to planning expertise involves some clear choices. Either one can adopt sustainable development as a policy goal or one can see it as altering the nature of the policy process. Many green commentators and planners hope that the two go together, that a more open and participatory policy process along the lines of LA21 will result in greener decisions. Environmental education is cited as a way of contributing to this synergy of goal and process, as is more open access to environmental groups and better resourcing of environmental arguments. Thus Healey also hopes to achieve both goals:

> Management by argumentation ... is premised on the belief that, through forms of inter discursive reasoning, democratic debate is both possible and desirable in contemporary pluralist societies. In the long run, such processes are also *probably* more efficient in achieving multiple objectives in the management of spatially differentiated change in urban regions, because they will be more informed and more capable of discovering lasting bases, *if any*, for consensus on strategy.
>
> (1995: 269–70, emphasis added)

But this synergy is not assured; there is no guarantee that environmentally beneficial policy outcomes will result from a more open process. This may not matter if either one has a greater commitment to altering the policy process than achieving specified outcomes in terms of environmental and spatial change, or if one is willing to wait and see if synergy will occur.

However, if one is of the view that there is a great urgency in the environmental agenda, then a commitment to the goal of sustainable development will dominate and this implies a different kind of policy process, a return to a more rational mode and the rise of the environmental technocrat. This tendency can be seen in contemporary environmental policy. European Union policy is largely of this type with the emphasis on achieving standards for environmental quality and utilising 'best available technology' in pollution control. The preference of central government for using the planning system to achieve climate-change policy objectives through manipulating urban form can also be seen in this light (Rydin 1995). Healey (1995: 257) refers to this type of planning as 'classically technicist' and also refers to the sustainable development agenda as leading to 'a return to tradition' (1995: 263).

The profession of planning is at a turning point. It has been subjected to wide and exhaustive criticism and has now, in effect, retreated to the status of managing the quasi-judicial process of town and country planning legislation. The process of environmental planning, as promoted by the Town and Country Planning Association (and as outlined in the previous chapter) will require a different range of skills and attitudes. The new environmental agenda is attractive to the town planning profession because it seems to hold out the prospect of a new legitimation. But the profession of planning, by its very nature, cannot provide all the approaches and vision necessary for the post-Rio agenda. The call for a 'new professionalism' is understandable but, in our view, naive – professions are structured within existing societal power relations – they have a role and function that cannot be subverted in this way. Similarly, to privilege the notion of communicative planning is simply to recognise the significance of politics.

Rather, the argument should be that the new environmental agenda of sustainability requires new approaches, new ways of working and a new politics. Knowledge can no longer be appropriated and designated as the domain of experts and professionals. The 'we know best' implicit in much professionalised planning of the post-war period does not sit easily with the post-Rio rhetoric of empowerment, capacity building and partnership. Planners, as professionals, need to recognise their position alongside other professions and groups within civil society. They need to be part, but only part, of an holistic approach to policy. They should hone their specific skills of administrative competence and urban design in the context of sustainability policy goals. And they should consider how they may contribute to environmental knowledge through their spatial understanding and everyday experience of planning practice in collaboration with others.

NOTE

1 Although we recognise the distinctions that may be made between the terms 'sustainability' and 'sustainable development', throughout this chapter the terms are used interchangeably.

5

PLANNING *IN* THE FUTURE OR PLANNING *OF* THE FUTURE?

Eric Reade

WHY THE READER MAY DISAGREE WITH EVERYTHING I SAY

The way we see anything depends, as we all know, on the concepts that we use to examine it with. In the first section of this chapter I therefore lay bare my conceptual apparatus. Before doing so, however, two preliminary explanations: first, I do not discuss here the way in which I think environmental planning *will* develop in the future, as I doubt whether, in terms of social science method, such prediction is possible. Instead, I discuss the way in which I think such planning *ought* to develop in the future. Second, these ideas about the way in which environmental planning ought to develop are expressed only *in broad outline*. Probably every single sentence in this chapter, therefore, really needs thoroughgoing elaboration and very careful qualification. For obvious reasons of space, however, I clearly cannot provide these things. The chapter should therefore be understood as intended only to provoke discussion, and not as a detailed examination of the issues involved.

In every society there is a dominant ideology. There must be, for without this ideological support, the ruling class could not rule. What every ruling class does, however, is to persuade the population that the ideology that legitimates its rule is not an ideology at all, but a set of factual statements about the objective nature of the world. It is *the others*, they tell us (i.e. those without power, and who want it) who peddle ideology. *They themselves*, our rulers tell us, are pragmatists and realists, their policies and actions reflect nothing but what, given the facts of the situation, common sense dictates, and in fact no reasonable human beings could do other than they are doing.

The truth is, however, that ideology is often more crucial to maintaining power than to challenging it. To legitimate an entire social system, it is necessary to find justifications for more or less every feature of that system. To challenge an existing regime, by contrast, no more may be needed than to show that it is failing in terms of its own definition of failure. There may be no need to construct an alternative ideology, which argues that the success or failure of society should be assessed by new, and quite other, criteria.

In the present political climate in Britain, for example, those who challenge the government find no need to argue that society ought to be assessed in terms of the extent to which it enables all its members to be creative and to achieve personal fulfillment. On the contrary, they usually find it sufficient simply to demonstrate that the present system is not meeting even its own crude and narrow criteria, which define national 'success' mainly in terms of GNP, interest rates, inflation rate, 'competitiveness' with other national economies, and so on. Those who challenge the existing system, in other words, often find it convenient to subscribe to most of the ideology that legitimates that system. It can serve their needs well enough. This is one of the reasons why dominant ideologies are so widely accepted, so pervasive and, indeed, so dominant.

Such dominant ideologies always include a judgement on the past, as well as a justification of the future state of affairs to which present policies seem to be leading. Where the dominant ideology is right-wing, it portrays these future states of affairs as desirable and anyway inevitable because they are in accordance with what it calls 'human nature', which it regards as fixed. Left-wing ideologies, dominant or not, usually portray their preferred states of affairs as *possible* rather than as inevitable, and as possible only as a result of encouraging the development of such 'higher' human qualities as compassion, co-operation and creativity.

The multitude of changes in society which together constitute the difference between the past and the future are referred to by sociologists non-evaluatively as 'social change'. In popular speech, by contrast, the same processes of change used to be summed up as 'progress'. This word often led to muddled thought, however, for it reflected a failure to make two necessary distinctions. First, it confused changes that happened by force of circumstances with those that were 'decided upon' in some way. Second, it failed to distinguish between changes seen as desirable and those seen as undesirable. Its widespread use as a synonym for social change therefore had the unfortunate effect of distracting our attention both from the question of whether we thought the changes going on around us needed to have happened, and from the matter of whether we welcomed them.

Those in power seldom nowadays speak of 'progress' anyway. Instead, they usually employ the sociologists' term, 'social change'. In doing so, however, they provide no improvement. On the contrary, they increase the confusion still further, for they misuse this sociological concept, and in at least two ways. First, most of what they call 'social change' is not social change at all, but technological change. Social change is not change in the technology available to us, but change in the way we treat each other. Second, wherever they can lead us to see the changes that are occurring as desirable, they accord to themselves credit for having produced these, or at least for having facilitated them. In the case of those changes for which they clearly *could not* by any stretch of the imagination take credit, they adopt the alternative ploy of inviting us to congratulate them on the rapidity with which they have adjusted to them, and on their having harnessed these changes in the world around us to something which they call 'the national interest'. The broad result is that the general public now perceive 'social change' in much the same way as they once regarded 'progress'. They see it as inevitable and anyway broadly desirable (or perhaps as desirable and anyway more or less inevitable).

This popular perception of things used to be summed up in the expression 'You can't stop progress.' Today, instead, it is usually summed up in the popular phrase 'The world has moved on.' And this phrase is a key part of the dominant ideology. By using it, we imply that those who regret any specific changes in society, and those who believe that any specific changes need not have happened, are out of touch with reality.

In this chapter, I challenge this perception. I suggest that although, as a central component of the dominant ideology, it is shared by most of us whatever our social situation or political convictions, this way of seeing things is promoted by and serves the interests of the rulers rather than the ruled. I also argue that this perception of change is factually untrue. I suggest, by contrast, that many if not most of the changes we see occurring around us were not inevitable, and are in many cases undesirable in human terms. Moreover, these changes, the most discussed of which are usually technological changes rather than social changes, do not reflect the workings of impersonal 'forces'. On the contrary, each one of them occurred primarily because some powerful interest group wanted it, and made sure it happened. The reason why the dominant ideology portrays most of the changes we see around us as desirable and anyway inevitable, I suggest, is that it is usually the powerful who get most benefit from them. It was the fact of being powerful, after all, that enabled them to influence the direction of change in the first place.

Broadly speaking, the expression 'The world has moved on' is therefore a statement popularized by the ruling class and summarizing that class's own 'achievements'. When the rest of us use this phrase, we unconsciously express approval of the way the powerful have used their power. For the changes occurring around us are not in the main *our* changes. When were *we* asked what changes we wanted? To the extent that the non-powerful influence the direction of change, they do so only at the cost of expending the most prodigious amounts of intellectual, emotional and physical effort, only a tiny part of which escapes being rendered futile by the powerful.

The ideology that legitimates both the existing state of affairs and the changes that are occurring, I term the 'dominant ideology'. This is because it literally dominates, and indeed, is virtually all-pervasive. I have already mentioned the case where 'opposition' groups find it unnecessary to construct an alternative ideology because they can oppose merely by pointing out that those in power have failed by their own criteria. But even where opposition groups *do* see themselves and *do* portray themselves as having an alternative ideology, analysis of this often reveals that, while it may at first *appear* to be different, it in fact rests on the same taken-for-granted assumptions as does the dominant ideology itself.

The dominant ideology in any society is therefore all-pervasive in a very strong sense of this term. It thus does not change easily or quickly. The mechanisms through which a dominant ideology *does* nevertheless change are not easily invoked, and may sometimes even require cataclysmic events to set them in motion. This explains why, for example, most of the citizens of the Nazi and Soviet systems supported those systems, if only passively, as long as they existed. Equally, it explains why, for example, even the strongest British Labour Party criticisms of the present Conservative government nevertheless reflect the same assumptions about what is possible and what is desirable as that government's own policies and actions.

The dominant ideology, then, is shared to a greater or lesser extent by virtually all. But it does not serve the interests of all. It serves the interests of the ruling class.

THE WORLD WE ARE TOLD WE LIVE IN

The present-day dominant ideology in Britain, I term Thatcherism. I reject the idea that Thatcherism began with Thatcher's accession or ended with her demise. On the contrary, I see Thatcher as no more than a vociferous mouthpiece for this ideology, which evolved some years before her arrival on the scene and which has grown even more dominant since her exit. I see the Major governments as more Thatcherist than the Thatcher ones, and regard the ideas currently set out by the British Labour Party as also resting on Thatcherist values. Thatcherism slowly supplanted the previously dominant ideology, that of the 'post-war consensus', over the second half of the 1970s.

The purpose of this chapter is to suggest how environmental planning in Britain ought to develop in the future. For reasons that will become apparent, this involves asking how Thatcherism might be supplanted, just as the 'post-war consensus' was. It is obvious that this could only happen if Thatcherism were to be seriously challenged. But to challenge an ideology we must first analyse it and dissect it. Among other reasons, this is because ideologies consist of four types of component, with quite different epistemological statuses. They consist of statements that are true, statements that are untrue, statements that are so general that their truth cannot be tested, and expressions of values, to which the concepts of truth and untruth are irrelevant.

The potency of any ideology depends precisely on its conflating these four elements, and on its weaving them all into a rather impenetrable seamless web of assertions. This is why most of us, not being trained in conceptual analysis, tend to 'buy the whole package'. This is how ideologies achieve the rather amazing effects that they do. I will not identify examples of these four types of component in Thatcherism, since I feel sure readers can do this for themselves. Instead, I will examine just four assertions, all central to Thatcherism, and all of which, I suggest, fall into my second 'type of component'. That is to say they are assertions purporting to be factual statements, but whose factual truth, I shall argue, is doubtful.

The first of the four may be summarized as asserting *that the economic problems that have beset Britain over recent decades have equally beset all the 'developed' countries, and are the result of 'economic forces'.*

The truth, I would suggest, is very different. To some extent, of course, it is true that recent economic depression has had international consequences. But Britain's current economic problems are largely peculiar to Britain, and are the consequence of the British government's extremist policies. I use the word 'extremist' here in a careful and specific sense. I use it to describe policies based on the assertion that a single and rather simple cause can account for a very wide-ranging and indeed virtually all-embracing economic and social outcome, and on a doctrinaire refusal to consider that other causes might contribute to that outcome. I also use this word 'extremist' to describe policies that are persisted in even when they manifestly do not work; Thatcherist governments have persisted in imposing policies that do not work because Thatcherist doctrine says they *ought* to work, and because Thatcherism is an ideology that places more faith in its own doctrines than in experience or evidence.

An obvious example is the doctrine which asserts that use of the market mechanism is the best means of solving virtually any economic problem. Running a close second to it is the assertion that financial gain is virtually the only psychological motivation that can induce us to contribute to the economy.

As regards each of these two assertions, the truth is very different. The market is indeed a most efficient mechanism, and can provide us with by far the best solution to many of the problems of production and distribution with which society is faced. But it cannot provide all the solutions. Some problems – and finding the most beneficial use of the earth's surface is a good example – are exacerbated rather than solved by applying market mechanisms to them. As for the problem of motivating us to contribute to the economy rather than allowing ourselves to become a burden to society, psychology can point to many such motivations other than desire for financial gain. These include the desire to achieve power or influence through our work, the love of creativity for its own sake, pride in doing as well as we can whatever we are set to do, the pleasure of co-operating with others in a shared task, and the sheer joy of being able to improve the lives and increase the happiness of ourselves or others without any material reward whatsoever.

Both these Thatcherist assertions – that the market alone can solve all economic problems, and that financial gain provides virtually the only way of inducing us to contribute to the economy – are crude, ignorant, puerile and simple-minded. The consequences of policies resting on the assumption that these doctrinaire beliefs are scientific truths are therefore unfortunate. Yet in Britain since 1979, whenever social scientific evidence concerning the workings of the market or the harnessing of human motivations has been necessary in framing legislation and policies, this necessity has been ignored. Instead, legislation and policy has been based almost entirely on these two simple-minded doctrines.

By any criterion, this is political extremism. Do we really need to ask why its result has been to produce or exacerbate social misery among the have-nots, and profound disaffection among all those with respect for evidence?

From 1979 to the present day (1996), then, Britain has been ruled by so-called 'conservative' governments which have been conservative in no sense, but which on the contrary can only be termed dogmatic, doctrinaire, market-obsessed, and pervaded by a naive conviction that economic outcomes have single and simple causes – in short, by governments of the extreme right. These have not been governments of conservatives, but of zealots and bigots. This is attested, for example, by the fact that pre-1979 British Conservatives, such as Edward Heath, have disowned these Thatcherist governments. It is also illustrated by the very considerable difficulties that members of these Thatcherist governments experience in working with Continental conservatives, supposedly their political soulmates, in the European Union. For outside Britain, conservative parties usually *are* conservative parties.

By contrast, since 1979 the British 'Conservative' Party has been a conservative party only in name; it is now an extreme-right-wing populist party. Conservatives in the true sense institute change carefully and thoughtfully. British governments since 1979 have, by contrast, consistently demonstrated a rash enthusiasm to introduce constant legislative changes for purely ideological reasons, and without pausing to enquire into the effects that all these legislative changes are having.

Only a society with a first-past-the-post electoral system, I suggest, could have produced such political extremism. In the more democratic constitutions and institutions of other west European countries, compromise ensures moderation. None of these British 'conservative' governments rested on a majority of the votes cast. Yet this strange British first-past-the-post system permitted these 'conservative' governments to engage in a seventeen-year-long propaganda-led campaign, designed to alter the very perceptions of reality on which society rests. Given that the British press is largely right-wing controlled (Hollingsworth 1986) and has contributed significantly to this campaign, this period gave more than enough time for the strategy to succeed.

These 'conservative' governments, in other words, have imposed the dominant ideology on society in a way and to an extent not previously known in this country. The constitutions of virtually all other west European countries, since they ensure that power is shared, would not have permitted such a programme of indoctrination. And largely *because* they have not been subjected to such a sustained barrage of extremist right-wing and free-market propaganda, they have been able to maintain more moderate and more socially minded economic policies, enabling them better to withstand the effects of international economic recession.

The belief widely shared in Britain that Britain's economic troubles have been equally experienced by the whole Western world, then, is itself merely British government propaganda, designed to conceal the fact that Thatcherism is the main cause of the economic ills of this country.

The second assertion of Thatcherism whose truth I would question may be summarized as suggesting *that the economy will of necessity become increasingly 'globalized', and that we have no alternative than to go along with this economic 'globalization'.*

The most striking feature of this assertion is its vagueness. Do the Thatcherists mean by it that the proportion of world trade accounted for by transnational companies is increasing? Or that some of these transnational enterprises are more powerful than certain national governments? Or that they are more powerful than *any* national government? Or that economic power in the world is increasingly monopolized by a small number of such transnational enterprises? Or do they mean that the countries in which the major transnational enterprises are based are using these enterprises as a means of promoting their own national economic and political ambitions? Or does this vague notion of 'globalization' refer primarily to the growth of international 'futures' markets in raw materials, currencies and so on? Or does it perhaps merely reflect the fact that the volume of goods and services traded across international boundaries is increasing? Perhaps those who talk so incessantly of 'globalization' mean merely that they *would like* any or all of the above things to happen, or that they welcome their happening. For, truly, it is very difficult to see whether these 'globalists' are identifying 'globalization', or advocating it!

If this assertion of 'globalization' suggests that economic power in the world is increasingly concentrated in relatively few hands, then its truth may certainly be doubted. Could it really be true, we might ask, that economic power in the world is more monolithic today than it was in the days when Britain was the only industrial nation in the world, when Britain was the only power exercising military might on a world scale, when London was the financial centre of the world, and when the British not only owned a vast empire consisting of a large part of the earth's land surface, but also 'ruled the waves' into the bargain? On the face of it, this seems rather unlikely. Could it be, we might hypothesize, that those who today talk so incessantly of 'globalization' suffer from a kind of blindness – which rendered them unable to see 'globalization' when their own country controlled the 'globe', and only able to see it when the centres of power moved elsewhere?

But while it is unclear what this Thatcherist assertion of 'globalization' is actually saying, what is by contrast very clear is that the Thatcherists *want* us to believe in the reality of this 'globalization'. For they assert it repeatedly and emphatically. But *why* are they so very keen that we should see this 'globalization' as so inevitable? Do they perhaps believe that their own power would be increased if, instead of continuing to operate within a relatively geographically restricted area, they were to become part of something bigger and more powerful? Or do they perhaps just want us to get used to the idea that we 'have no alternative' than to work in Japanese-owned factories? The most credible hypothesis, I would suggest, is that this slogan of 'globalization' (for in truth it is no more than a vague slogan) serves as a justification for governmental inactivity under Thatcherism. It would be futile for government to attempt to shape the way the British economy develops, Thatcherism asserts, because the 'fact' is that an 'economic force', 'globalization' is inexorably shaping the development of all the national economies in the world. And being an 'economic force', or a product of 'history', 'globalization' cannot be resisted.

We could not carry out research to establish whether 'globalization' 'is' or 'is not' happening, I would suggest, since this concept is too vague to be operationalized. It is merely a slogan, serving an ideology, and is not a usable social-scientific concept. We would be better advised, I think, to ignore it, and to concentrate our attention instead on two highly undesirable developments in the international economy which without any doubt are happening, and which, in contrast to the nebulous notion of 'globalization', are perfectly concrete, graspable and observable.

The first of these is the international exploitation of people in poor countries. The second is international speculation in commodity and currency markets, which destabilizes and weakens the weaker national economies. For brevity, we can label them the exploitation problem and the speculation problem.

Looking first at the exploitation problem, we see that, increasingly, raw materials, semi-finished products and product components are transported over vast distances in order to be refined, processed or manufactured in low-wage economies, and the refined, semi-finished or finished products are then transported over vast distances to be sold in high-wage economies. Capitalism has always worked in this way, getting things made where wages are low, and selling them where people can afford to pay more for them because wages are high. Reductions in the real costs of international transport, however, have made it far easier for this to be done on a vast geographical scale. And because by international standards the differences in the wages for which people in various parts of the world are prepared to work are truly staggering, the opportunities for such exploitation have, with cheaper transport, expanded enormously.

Increasingly, therefore, the capitalist system today divides the world's population into those, on the one hand, whom it persuades can achieve self-esteem only by consuming cheaply produced needless luxuries, and those, on the other hand, who can stay alive only at the cost of being mercilessly exploited in the production of these goods. And international capitalism lives and grows and prolongs its own existence by bringing these two groups increasingly into dependence upon each other.

It takes little thought, however, to see that this development is highly undesirable. And it takes only a little more thought to see that by acting jointly, people in rich and poor countries could oppose it and eventually end it. Once we know the scale of the exploitation involved, I suggest, we in the 'developed' countries simply will not wish to purchase goods, knowing them to have been processed or manufactured by people obliged to work inhumanly long hours in exchange for bare subsistence wages, for this must surely reduce our enjoyment of these goods. It is also becoming very apparent to us that owning too many goods can make our own lives as difficult as, only a very short time ago, they were made difficult by owning too few goods. We are beginning to understand how we can enlarge our personal freedom by living simpler lives, and this too is likely to reduce our appetite for a plethora of possessions.

Turning to what I have identified as the second undesirable feature of the present-day international economy, speculation in materials and currencies, the undesirability of this is quite as apparent as is the international exploitation of labour. This speculation is strictly speaking a *non*-economic activity, for it produces nothing, neither goods nor services. It serves the interests only of the speculators themselves, and not those of the rest of us, for it puts at risk the livelihoods of all who are in any way dependent on the materials or the currencies that are the objects of the speculation. This speculation also contributes to the processes whereby much-needed capital is being constantly drained away from poor countries towards rich ones. It seems unnecessary to elaborate. One of the strongest arguments for a single currency within the EU, for example, is that this would put an end to the harm done to us by those who affect the fortunes of our separate national economies by speculating in our national currencies.

Looking jointly, then, at these two specific problems of the international economy, exploitation and speculation, we see that there is little doubt that they could be ended. Despite what Thatcherists tell us, they result not from 'laws of economics' or 'laws of history', but from the anti-social actions of specific persons. History is not made by 'laws of history' or 'laws of economics'. On the contrary, history is made by human beings. The course of events does not run in a straight line in accordance with a predetermined pattern. It proceeds, rather, by action and reaction. What happens in the future, therefore, depends on *how* we react today to what we see happening around us.

Exploitation and speculation at the international level, then, should be opposed, just as exploitation of workers was opposed in the early stages of capitalist industrialization. And for the same reason: unrestrained, capitalism is barbaric. Wisely regulated and controlled, it can be a very useful, and arguably the only, means of achieving highly desirable social objectives. The excesses of unrestrained exploitation and speculation at the international level are in fact already producing their own reaction. Already, there are many all over the world working to persuade governments to adopt more economically rational alternatives. Already, many influential organizations are working to persuade governments that we should regard members of other societies as fellow human beings, and not as customers for needless luxuries on the one hand, or as exploitable labour on the other. And already, some governments – though not of course the British government – are heeding such arguments.

These undesirable developments in the international economy are also being checked by growing political pressure for 'sustainability', which among other things demonstrates the economic irrationality of devoting such a large proportion of our resources to transporting goods over vast distances merely in order to get them processed and used in manufacturing processes by people paid bare subsistence wages. I would therefore suggest that international exploitation and speculation will be checked by the increasing power of societies, and groups of like-minded societies, to assert their right to organize their economies in ways that promote their shared social values – to demonstrate that they can, after all, prevail over the international industrialists and financiers who put their own profits before human life (cf. Hirst and Thompson 1996).

The third assertion of Thatcherism whose truth I would question may be summarized as suggesting *that since 1979, 'the world has moved on', and that there can therefore be no return to the Keynesian economic policies or the 'welfare state' which we enjoyed before that date.*

I have suggested that the course of human events often runs not in a straight line but by action and reaction, and there could be few better illustrations of this than those provided by Thatcherism itself. Thatcherism shows us how those in power, reacting strongly against the way society has changed, can by sheer political will re-create the past. Virtually all that British governments have done since 1979 can be explained as a powerful and sustained reaction to the welfare state (Bruce 1973), and especially to the social gains enacted in the 1940s and further consolidated in the 1960s. It is important here, however, to remember my distinction between social change and technological change. I am certainly not suggesting that Thatcherism has sought to reverse the *technological* changes that have occurred since the Second World War. Certainly not, for these changes in technology are the lifeblood of capitalism, and often of a particularly 'raw' variety of international capitalism which is particularly congenial to Thatcherists. But Thatcherism *has* attempted, and indeed, has to a large extent succeeded, in reversing much of the *social* change that has occurred since the 1930s. And the term 'social change', here, includes *economic* change.

Social change, I suggested, is change in the way we treat each other. And here Thatcherism has succeeded only too well in turning back the clock. Since 1979, for example, those in power have persuaded us that employment is not a fundamental human right but a privilege conferred on us by employers, and it has succeeded in restoring to a large extent the unequal relationship that existed between employers and employees before the Second World War. It has persuaded us that trade unions should play no part in promoting our political and social education, or in government, but should have no function other than to bargain for wages. It has persuaded us that we owe few obligations to our fellow human beings, even those who are unfortunate, inadequate, oppressed, exploited or disadvantaged, and that their needs should be met only if they can pay for it. It has transformed local government from a forum in which we learn to be useful and responsible citizens into to a mere business enterprise, from which we buy services as we would from any other profit-seeking entrepreneur. It has made the manufacture and sale of armaments into one of Britain's very few thriving and successful industries, made it into a central element of the British economy, and persuaded us that if 'foreigners' are foolish enough to buy these weapons and kill each other with them, we should have no qualms since this provides us with jobs. It has encouraged the natural British talent for xenophobia to the point where it has become one of our chief national pastimes, and it has taught us that those who flee from political terror deserve little help because they are 'economic migrants' (which, translated into English, means that having escaped terror, these refugees hope to work in order to support themselves; Thatcherists apparently find this unreasonable).

These propaganda-led transformations are well documented (see, for example, Blunkett and Jackson 1987, Hirst 1989, Hutton 1995, International Broadcasting Trust 1994, James 1995, Kingdom 1992, Marsh and Rhodes 1992, Skidelsky 1988. The consequences for the planning system are discussed by, for example, Brindley *et al.* 1989, Montgomery and Thornley 1990, Thornley 1991, 1992).

Clearly, one could go on. In summary, Thatcherism has succeeded in persuading us that, in the way we interact with our fellow human beings, morals and sensibilities need play no part; it has persuaded us that in principle, and as far as possible in practice, all human interactions are best effected through the medium of money.

The results of Thatcherism, then, may be summed up as a return to the harsh social attitudes and values and economic arrangements of the 1930s – but at a vastly higher level of technology. Whereas the unemployed of the 1930s marched on London, those of today are socially atomized, and thus sit isolated in their own homes, watching videos. Thatcherism has persuaded them that they are not a class, which could improve both its own situation and the state of the world by collective action, but that they are merely inadequate individuals who don't deserve to buy goods and consume them at as fast a rate as those who have been more 'successful'.

In 1945, the phrase on everybody's lips was 'never again' (Hennessy 1992; see also, for example, Addison 1977, Calder 1968). Never again, we said, would we allow our society to return to the social and economic attitudes and values and practices that produced the harsh class divisions and the cruel social injustices that existed in the inter-war period. Yet this is precisely what we *have* allowed. The Thatcherist 'Back to the 1930s Project' has succeeded to the extent that Britain has been rendered into one of the most socially and economically backward countries in Europe. According to certain researchers at least, Britain's 'democracy rating' is lower than that of virtually any other Western society, and lower than those of Hungary and the Czech Republic (Smyth *et al.* 1994). Young Britons now move to Germany and Holland to work for low wages as those from Turkey or Spain once did, performing tasks that Germans or Dutch people would not even consider.

And all this has been achieved through the power of the ruling class to evolve and propagate a 'dominant ideology' that shapes our perceptions of the nature of reality. Since 1979, the dominance of Thatcherist ideas in society has re-shaped our beliefs as to what is right and wrong, desirable and undesirable, possible and impossible. Above all, it has re-shaped our beliefs about human nature.

The particular part of this re-shaping of perceptions with which I am concerned relates to politics. Here, Thatcherism has caused us to lower our sights to a remarkable degree. It simply is not possible, Thatcherism has persuaded us, for human beings to act collectively to shape the way society develops, for the fact is that the development of society is shaped by inexorable economic laws. To engage in public debate about the way our society ought to develop, therefore, is to engage in a futile activity. The only sphere in which we can effectively express our preferences, it tells us, is in the market, as consumers of goods and services.

None of what Thatcherism tells us, of course, is necessarily true. But social science enables us to understand why it is that, as with all social beliefs, it becomes factually or empirically true to the extent that we believe it to be true. But by the same token, social science equally explains to us how it ceases to be true if we decide it is untrue. And the only thing that would enable us thus to see these Thatcherist assertions as untrue would be our possession of an alternative ideology, providing us with a different perception of reality. Such an alternative ideology, for example, might see these Thatcherist assertions as pernicious and life-denying. I would suggest that, in persuading us that we have no power to shape our society and can therefore only find solace in shopping, Thatcherism *is* pernicious and life-denying.

The fourth assertion of Thatcherism I call into question may be summed up as suggesting *that due to the 'collapse of communism' there can be no resurgence of public demand for socialism or planning*.

Several years before the collapse of the USSR, its ambassador in Stockholm, Boris Pankin, who was clearly more frank than most, publicly remarked that Sweden was a far more socialist country than his own. I doubt whether Swedes saw this as more than a statement of the obvious. Interest centred not on what the man said, but on the fact that he dared to say it.

Those who believe that the former USSR was either socialist or communist would presumably believe anything. This assertion was formerly made only by the USSR's rulers, and in the West, for their own purposes, by the political right. By now, however, it has been repeated so often that it is believed even by most of those on the left. In reality, by contrast, the USSR was a militaristic police state, in which fear of consignment to a labour camp, or later, to a psychiatric ward, prevented nearly all freedom of expression. It was also profoundly inegalitarian, the elite and the masses having utterly different living standards. Though its rulers claimed to rule in the name of 'communism', or even, on occasion, 'socialism', the USSR was about as far removed from socialism or communism as it is possible to be.

Sweden, by contrast, is a society in which equalization of incomes has been taken to lengths undreamed of in the USSR. A very long period of almost unbroken social democratic government, and the consequent diffusion of its characteristic value system throughout all the political parties, has produced in Sweden a society in which most people have a sense of social solidarity, and in which social and political alienation is probably far more rare than in Britain. Consequently, Swedes have remained content to pay a far higher proportion of their incomes in tax than we do in Britain. They seem on the whole to believe that to spend their incomes in this way, to produce a more egalitarian and more humane society, is sensible. And contrary to what we in Britain are told, the present government in Sweden broadly maintains these arrangements, despite economies in spending neccessitated by economic recession. It was undoubtedly these attributes that Pankin had in mind when he called Sweden socialist (see, e.g., Anton 1975, Castles 1978, Heclo and Madsen 1987, Khakee *et al.* 1995, Montin and Elander 1995, Reade 1989b).

As for the Thatcherist assertion that there is no demand for socialism in today's world due to the 'collapse of communism', I could point out that, in the Swedish general election of September 1994, 45 per cent of the votes cast were for the Social Democrats, 6 per cent were for the former Communists and 5 per cent for the Greens, a combined left-wing vote of 56 per cent. Bearing in mind that the four 'right-wing' parties, the Moderates, the Centre Party, the Liberals and the Christian Democrats, would probably all appear to be far from right-wing in the eyes of most British people, and that the 'centre point' in Swedish politics is thus considerably further to the left than it is in Britain, this looks like a *rather strong* demand for socialism.

A VERY LITTLE HISTORY

The message of the previous section might be summed up as 'It ain't necessarily so!'; seen from a different perspective, the world we live in appears very different from the world as it is presented to us by the dominant ideology. Another way of freeing ourselves from the blinkers placed surreptitiously on our heads by the dominant ideology is by stepping out of the present and studying a little history. And we need do only a *very* little. We need to study only as much as is required, I suggest, to remind ourselves and to convince ourselves that the past was very different from the present. This will help us to believe that it is just possible that the future could be very different from the present, too.

I have argued above that four of the central assertions of Thatcherism are factually untrue. What is the relevance of this to the future of the British planning system? When the Attlee government established this system in 1947, it was seen as part of a wider system of economic and social planning (see, e.g., Ashworth 1954, Cherry 1974, Cox 1984, Hague 1984, Ravetz 1980, 1986, Reade 1987). Over subsequent years, governmental support for this conception of physical planning as part of a wider system of economic and social planning fluctuated. It weakened, for example, during the 1950s, but returned very strongly in the 1960s. It was not generally abandoned, however, until, in the second half of the 1970s, the dominant ideology that supported the 'post-war consensus' gave way to Thatcherism.

All four of the assertions that I have identified as central tenets of Thatcherism lean in one direction. They all tend to deny that we can collectively shape the way society develops, and they all suggest, by contrast, that the course of events is shaped by impersonal forces largely beyond our control. Thus, Thatcherism attributes the economic ills of Britain from 1979 to the present day to economic forces in the world at large, rather than to any actions or failures to act on the part of the British government, and it suggests that we can do little to resist the forces of 'globalization'. It asserts that we cannot choose to rescind the Thatcherists' own demolition of the welfare state because we are now in another 'stage of history' in which welfare states supposedly cannot exist. And it asserts that we cannot choose to introduce socialist measures since 'history' has shown socialism to be impossible in the present-day world.

What is striking about Thatcherism, then, is its fatalism. It asserts that the pattern of economic and social development is determined by historical inevitability and impersonal 'forces', and that we are therefore free to make our social arrangements only within very narrow limits. In this, Thatcherism closely resembles Marxism. Indeed, my own reading of Marxist thought indicates that it tends usually to accord to human beings rather *more* scope to 'make their own history' than does Thatcherism. I find these Thatcherist assertions of inevitability untenable. I think history is made by human beings, and not by impersonal 'forces'. Though it ascribes such weight to them, Thatcherism is vague in the extreme about the nature and the workings of these 'forces'. It does not, for example tell us *why* welfare states should be possible in one stage of history but not in another. Marxism, by contrast, and quite apart from whether one finds it convincing, does at least provide us with *reasons* for its assertion that societies must of necessity go through the various 'stages' of economic development it specifies.

I suggest that the Attlee government was correct in assuming that we can as a society collectively shape the pattern of physical, social and economic development, and that Thatcherism is wrong in asserting that we cannot. But this needs to be stated rather more carefully, taking account of the role of perceptions. More accurately, then, we should say that such collective shaping of the pattern of development tends to be impossible where it is perceived as impossible, but tends to be possible where it is perceived as both possible and desirable.

But while it is possible, such collective decision-making about the pattern of future development is difficult. First, and most obviously, it is difficult because before it can happen it is necessary to create a public perception of it as both possible and desirable. Another reason why it is difficult is that to do anything – even the simplest thing – jointly with others can be fraught with difficulty. To collectively shape the pattern of development of society as a whole, therefore, is a fairly ambitious undertaking. But I would deny that it is an impossible one. And no matter how difficult it may be, it seems essential that we attempt it. For if we do *not* do so, we shall cease to be civilized beings. In the nineteenth century, for example, it proved very difficult for workers to learn how to get together to form unions, and thus to act collectively to put an end to the barbaric exploitation practised by the early industrialists. But it was done, nevertheless. And it is in part because it was done that we enjoy today the level of civilization we do. Today, similarly, it will no doubt prove difficult for nations to learn how to act collectively to put an end to international exploitation of labour and international speculation in raw materials and currencies. But the way in which present-day transnational capitalists and speculators put their own profits before human life is akin to the attitudes of factory owners in the early years of the Industrial Revolution. The future of civilization therefore depends in exactly the same way on our collectively destroying their power to act in this way.

Under Thatcherism, the belief that physical planning should be part of wider economic and social planning has been abandoned. Central government has made it clear to local authorities that they will not be permitted to pursue economic or social aims through the use of physical planning powers. The legal fiction that planning is concerned only with physical arrangements has been enforced very literally and very thoroughly; a legal fiction that has become a legal fact, we might say.

My argument, however, is that the Attlee government's approach was well advised, and that we should return to it. For only by doing so can we again collectively take responsibility for the future development of society and its environmental arrangements. I shall term this taking of collective responsibility for the future development of society and its environmental arrangements 'effective planning'. But this effective planning will only become possible, I suggest, if there first emerges a more moderate political climate.

POLITICAL MODERATION AS AN INSTRUMENT OF RADICAL CHANGE

In this section, I shall argue that the creation of an effective system of environmental planning in Britain, like the achievement of other kinds of radical social change, depends on the emergence of a far more moderate political climate than presently exists in this country. This last, however, is best brought about not by exhortation, or by changes in attitudes, but by adopting constitutional arrangements that *compel* political moderation. I shall advance this argument through the following linked ideas: though all the other aims of Charter 88 are equally desirable and necessary, I would argue that probably the most important of that group's aims from the point of view of planning is proportional representation. This is because it would oblige governments to take account of a wider spectrum of informed opinion and, in particular, because it would strengthen the single-issue pressure groups, who are the best possible allies of those who work for an effective system of environmental planning. The more moderate political climate brought about by proportional representation would also make it possible to bring land taxation back onto the political agenda. And this is essential, for recoupment of development value is an absolute precondition of any effective planning system.

Above, and defining my meaning carefully, I have characterized the policies pursued by British governments since 1979 as 'extremist'. The impact of these policies, and perhaps even more so of the ideology by which these policies have been justified, has quite naturally been to produce a social climate that is itself socially and politically extreme; it is hard, lacking in compassion, and prone to label as 'utopian' any endeavours which seek to harness any human motives other than selfishness and love of financial gain. But neither the policies nor the resulting social climate are widely *seen* as extremist. Incessant governmental propaganda has persuaded the population that they are, on the contrary, merely unavoidable pragmatic responses to the 'realities' of the world we live in.

The first problem, then, lies in demonstrating just how extreme the present social and political climate in Britain is. The second problem lies in explaining what a more 'moderate' political climate would look like, and how it might be created. One good way to initiate the former would be by statistical evidence. This is hardly possible here, but if, for example, people were more aware that the income and wealth differentials between the richest and the poorest groups in society are greater today than they have been for over a hundred years – and of many other similarly striking facts about present-day Britain – then the extent of Thatcherist extremism would become apparent. All the facts concerning the economic decline, the social misery and the widespread alienation brought about by Thatcherist extremism could be similarly quantitatively demonstrated. Conversely, it could be pointed out that, by contrast, the period of Keynesian economic policies and relatively high social spending from 1945 to the mid-1970s produced the longest period of unbroken growth in enonomic and social well-being in our history – even though much of our national wealth over that period was going into defence spending.

To the observer, it would seem that if such facts were more widely known within Britain, public support both for radical socialism and for extensive social redistribution would probably be stronger today than at any time in the country's history. But such facts are apparently *not* widely known in Britain, and one reason for this is that the British Labour Party has not based its policies on making such facts more widely known. Instead, it has chosen simply to occupy the ground that the Conservative Party has vacated. The Conservative Party is no longer conservative, but has moved away to become an ultra-right-wing populist party, while the old British Conservative tradition has been inherited by the Labour Party. This pronounced rightward shift on the part of Labour has the unfortunate effect of producing a political climate even more extreme than Thatcherism alone could have produced, for it means that the *fulcrum* of British political life has moved far to the right. This could be seen as alarming.

It may seem rather less alarming, however, when we remember that over the last three decades or so, by far the greater part of such enlightened social reform as has been achieved in Britain has been brought about not by the political parties at all, but by single-interest campaigning groups. There would be no exaggeration, I suggest, in regarding these campaigning groups as currently providing a better guarantee of democracy than does the existence of either the political parties or of parliament. The latter, after all, only respond, whereas the campaigning groups initiate. And while in relation to any specific area of policy the institutions of the formal political system are active only from time to time, the campaigning groups are alert and active all the time, each keeping events within its own sphere of competence under constant surveillance. These campaigning groups not only engage in direct consultations with civil servants, but, unlike politicians, also make the contents of these consultations public. In engaging in public debate, they provide us with very high levels of information. Their debating style is adult, and their very real concern for their subject matter is manifest. They therefore command widespread respect. Politicians, by contrast, increasingly marginalize themselves as a group because they choose to operate at very low intellectual and moral levels, and to collude with the media in the personalization and the trivialization of those issues that they jointly and tacitly agree from time to time shall be publicly ventilated.

Those working seriously for any social reforms, therefore, are more likely to elicit a well-informed response at the appropriate technical level if they make their appeal to the campaigning groups rather than to the political parties. And some of the most successful campaigning groups in Britain –

especially, perhaps, those in the environmental field – have much the same concerns as do 'enlightened' or 'politically conscious' planners. They have much the same values, and seek to promote much the same causes. Their prospects of achieving jointly with planners an effective environmental planning system in Britain would therefore seem to be good.

I would suggest that, of all the various pressure groups with which the advocates of a more effective planning system might join forces, the most important are those campaigning for the taxation of development value. This, after all, was seen as a precondition of effective physical planning by the Labour administrations of 1945, 1964 and 1974 alike, and there is no doubt that in this they were correct. All three of these legislative attempts to recoup development value were of course repealed at the first opportunity by the three respective incoming Conservative administrations. The history of land-value taxation in Britain over the post-war period therefore provides an excellent illustration of the way in which the British first-past-the-post system inevitably produces violent swings in public policy.

When British planners discuss this matter of why all these three Labour attempts to tax land values 'failed', and why neither Labour nor the Liberal Democrats presently advocate a fourth attempt, it is sometimes suggested that this is because such recoupment of development value is inherently difficult. This hypothesis, I would suggest, can be immediately rejected; most developed countries, after all, impose such taxation of development value in some way or another, for without it, their attempts to control the pattern of physical development through their statutory planning systems would be largely unworkable. A far better explanation of the present-day absence of land-value taxation in Britain, I therefore suggest, lies in the nature of British political institutions, with their inbuilt proclivity to produce political extremism.

In the British first-past-the-post system, with its unfortunate adversarial and polarizing tendencies, there is an inevitable tendency for the party taking office, however slender its majority, to legislate at breakneck speed, without either the necessary period of public education in the matters at issue, or the necessary consultation with the knowledgeable interest groups concerned. As a consequence, we have in Britain a silly see-saw system, in which one party enacts its less-than-properly-thought-out schemes as soon after taking office as it can, and the other party, equally mindlessly, repeals these measures the moment it gets back into power.

Again, Sweden provides an alternative model, though one which is found to a greater or lesser degree in all other West European societies. It is rare in Sweden for legislation to be repealed. This is because while it is in preparation, all conceivable experts and interest groups, including opposition parties, are consulted about it, and because this process is not a mere formality, but a real process of negotiation and exchange of knowledge. All such consultees, we might thus say, are thus 'hostages'; they can hardly subsequently object to measures to whose formulation they have themselves contributed (Anton 1975). Nor should we believe, though this is what we in Britain are constantly told, that such a system leads to weak compromises. While it is certainly true that in this system all interests are taken into account, it should not be supposed that all are *equally* taken into account. On the contrary, they are taken into account in proportion to their actual power and persuasiveness. One political scientist provides a good summary: in a first-past-the-post system, he remarks, the winners take all. In one with proportional representation, the winners take their due share (*ibid.*).

Slow, patient formulation of policies, as opposed to what I have called a silly see-saw of constant enactment and repeal, is an integral part of political systems based on proportional representation. The current Swedish planning act, for example, which dates from 1987, was in preparation for about fourteen years before becoming law, during which time consultations proceeded with all concerned groups. It is therefore a true reflection of the extent to which ideas had changed, among all concerned groups, since the previous act. Perhaps the most important point emerging, I would however suggest, is that such a slow, careful, patient and conciliatory approach, unlike the adversarial British one, permits and encourages public education in the issues at stake. The real reason why, despite three legislative attempts, we still have in Britain no recoupment of development value, I would suggest, is that our political institutions do not permit or encourage either public debate of the issues at stake, or public education concerning them.

The first-past-the-post electoral system and the legislative see-saws that are its inevitable outcome, then, produce trivial levels of political debate. We see this constantly in the British media, where even in the 'quality' press and in 'serious' television presentations, the level of analysis is relatively shallow and sensational. As long as we have this crude 'first-past-the-post' electoral system, there is little incentive for us as planners to engage in the considerable work involved in raising public understanding of the problem of land values, or in persuading the Labour Party once again to tax development value. For we know that the resulting adversarial political culture, with its rowdy 'public school dormitory' parliamentary style, will virtually guarantee both that any legislation will be inadequately thought out, and that it will be mindlessly repealed at the first opportunity anyway.

Had we the more democratic and more mature political institutions based on proportional representation that permit and encourage more rational public debate and more thorough public understanding of what is at issue, I feel sure that we would have long ago adopted a system of taxation of land values sufficiently well thought out to have become permanent. The case for it is overwhelming, and all would understand this given the chance. (The same is true, of course, of many other much-needed social reforms in Britain.)

DEFINITION OF THE PROBLEM

The 'problem', I suggest, is that what we have now does not constitute an effective system of environmental planning as we understand this phrase. But this statement requires a little elaboration. If it is denied any real function in the national economy as a result of the absence of taxation of development value in land, any system of physical planning must inevitably become merely cosmetic, symbolic and to a greater or lesser degree a sham. And in the case of the existing British planning system, this is exactly what we see. The British planning system has become a mere 'token' system. It 'works' by portraying itself as safeguarding the public interest in land and property development against private and sectional interests. But by means of its assertion of the legal fiction that planning has no economic or social objectives, it prevents all public knowledge of how this assertion of public against private interests is achieved. It does not tell us how 'public interest' is defined, or *why*, exactly, the forms of development and the spatial patterns that emerge promote the public interest. From the viewpoint of planners themselves, I suggest, reliance on this legal fiction does not help. The doctrine which insists that local planning authorities, in deciding whether to grant permission for proposed development, must take no account of its social or economic effects, but must on the contrary assess it only in terms of whether it constitutes 'good planning', explains nothing. On the contrary, it is a form of mumbo-jumbo which probably serves only to raise public suspicions.

Such suspicions are probably justified. The social function of the planning system, I suggest, is to provide a facade or smokescreen. The bland but totally unverifiable assertion that both in any given case, and as regards the workings of the system as a whole, the public interest 'is' being safeguarded against private or sectional interests makes it *more* difficult for us to see what is actually happening in terms of losses and gains than if we had no planning system at all. For this system excludes consideration of losses and gains from polite conversation. We must only discuss 'good planning', and 'amenity'.

There is little point in retaining this token system. It is difficult to discern any public benefits that it may be producing. Indeed, it is difficult to detect any *effects* that it may be producing, beneficial or otherwise. The last and virtually only attempt to research the effects of the British planning system suggested that it served generally to increase the environmental and financial advantages of the affluent and the privileged, living in places socially perceived as desirable, and to decrease those of the less affluent, living in less desirable places (Hall *et al.* 1973).

Although evidence is lacking concerning the social benefits of this system, it nevertheless seems clear that it produces private benefits, at least for some. Its concealment of the distributional consequences of property development, and its bland assurance that, because the patterns of development that emerge are 'planned', we can rest assured that they promote the public interest, must surely be very convenient to those involved in the property-development industry. It gives them respectability. Indeed, it puts their operations virtually beyond scrutiny. And though this existing planning system provides relatively well-paid employment for a small army of planning consultants, these latter never seem to bring us very much closer to any very real or very widespread public understanding of what this planning system is all about. To most ordinary people, I suggest, 'the planning' remains largely a mystery. As to how exactly we are all made better off as a result of developers' need to 'get planning' before they can proceed, this is surely something that few ordinary citizens could explain.

In the next section, I discuss the matter of moving from this 'token' planning system to an effective planning system. By 'an effective system' I mean one whose aims are clear and widely understood, and in which progress or lack of progress towards the achievement of those aims could be readily observed by ordinary people.

NINE THESES ON THE CREATION OF AN EFFECTIVE SYSTEM OF ENVIRONMENTAL PLANNING

I set out below nine statements about this matter of moving from what I have called a 'token' planning system to what I have called an 'effective' one. They are not intended as a political programme, but as social scientific statements about causal relationships. I am tempted to remark that to me, they are simply self-evident truths. But lest there be those who do *not* see their truth as immediately apparent, I will call them theses, whose truth might be discussed.

1. *An effective planning system must logically rest on effective mechanisms for the recoupment of development value in land.*

One reason why expropriation of development value is essential to planning is that real planning, as opposed to the present system of token planning, has a cost. Local planning authorities would wish, for example, to provide open space where the market would provide none, and to conserve buildings which the market would not conserve. They would want to discourage the emergence of single-class 'ghettos' by enabling people to live in places where, in a free market, they could not afford to live (cf. Donnison and Soto 1980), as well as to provide local employment and a proper level of social facilities. They would also wish to raise environmental standards in all manner of ways which powerful interests, accustomed to operating in an unregulated free market, would ensure were publicly perceived as 'utopian' and denigrated as 'uneconomic'. Since nothing is costless, such authorities would need an income with which to pay for these social goods. Even where, as I would argue, no actual payment should be made, as where land is simply zoned for a less profitable use than the market would ascribe to it, there is nevertheless a cost to society, whether in terms of developers' profit (and thus tax) foregone, or (arguably) in terms of departure from pure economic efficiency. But beyond this 'pragmatic' reason for expropriating development value – that we simply need the money in order to pay for a civilized environment – there are two other reasons for this expropriation. We may call these the 'ethical' and the 'economic efficiency' justifications.

The ethical justification is often stated, and need be mentioned only briefly. It rests of course on the fact that the development value in land is socially created, rather than resulting from the efforts of the owner or the developer, and that it is therefore socially unjust *not* to return it to the community.

The economic-efficiency argument points out that where development value is *not* expropriated, those who produce buildings, unlike those who produce prams or sewing machines or ice cream or virtually anything else, make a large proportion of their profits not out of their skill in organizing the production of these buildings, but out of the increase in the value of the land on which they construct them. This being so, they can usually make very comfortable profits without taking the trouble to organize that construction process very efficiently. In Sweden, to continue with the example I have used above, the economic efficiency of the building industry was thus considerably increased as a result of this expropriation of development value. Builders were forced, like those who produced prams or sewing machines or ice cream, to make their profits by their skill in organizing the construction process, and not by simply waiting for the land on which they had been granted permission to build to increase in value. This forced them to be more economically efficient. By 1980, the construction costs in real terms of small houses in Sweden were only 98 per cent of what they had been in 1965, while building labour productivity rose by almost 250 per cent from 1950 to 1980 (Duncan 1985: 331; for discussions of land-value taxation, see, e.g., Cox 1984, Dickens *et al.* 1985, Douglas 1976, Douglas 1993, Hall 1965, Hallett 1977 and 1979, McKay and Cox 1979: ch. 3, Reade 1987).

In thus summarizing the 'economic efficiency' argument for expropriation of development value, it should also be remembered that there is probably no other form of taxation that has the effect of increasing economic efficiency rather than of decreasing it by creating 'market distortions'.

As to *how* development value might be recouped, this is also well-trodden ground, and needs little repetition here. Basically, the two main methods are either by requiring all development land to go through the hands of a public body which buys at existing use value and sells at development value, or by imposing a periodic tax on all increases in land value. The assertion that it can be done through 'planning gain' and 'planning obligations' I find ludicrous. The effect would be merely to invite corruption; this would be (in fact *is*) an optional tax, raised by horse-trading, and in secret (cf. Ward 1993).

2. *In order to move from the present obscure token planning system to one that is effective in the sense that it produces desired and comprehensible outcomes, it will be necessary to introduce a far greater measure of openness and democracy into that system.*

Given its existing 'social function' of concealing and legitimating the operations of the development industry as identified above, it is hardly surprising that one of the most striking features of the present planning system is its tendency both to secrecy and to intimidation of the public.

Imagine for a moment the impressions of an ordinary citizen wishing to take part in a public local inquiry into a planning appeal. One enters the local council chamber where highly paid barristers are engaged in learned and rarefied disputation in their efforts to promote the interests of some powerful national developer. The impression one gets is that the matter is so far above one's head that one would feel quite foolish in speaking. Yet the fact is that these are administrative proceedings and not judicial ones. It is therefore quite improper that the legal profession should ever have been permitted to 'judicialize' them and to impose its own professional procedures upon them in this way.

Increased openness and democracy in planning, I suggest, will be achieved only if the proponents of planning join forces with those campaigning for constitutional reform. To help produce a more open and more democratic climate, many changes would be necessary. By way of illustration, I can mention just a few of the more obvious ones.

It would first be necessary, as I have suggested, to put an end to the system in which the legal profession imposes its own intimidating medieval rituals and procedures, mimicking those of courts of law, on planning inquiries. To say that the public find this intimidating is an understatement; in effect they are prevented from participating fully and effectively in them, despite the fact that they have every right to do so. These are supposedly *public* occasions, and their purpose is supposedly to facilitate public involvement, not to inhibit it. If the planning system is to be open and democratic, this fact would have to be reflected in the way they are conducted. Members of the public could, for example, be provided with the same desks and other facilities as are the paid professionals taking part. To emphasize that these are simply inquiries into questions of public concern, all participants could be seated in a circle and at the same level, and be relieved of any need to observe such undemocratic rituals as standing to speak, or being obliged to address the chair as 'Sir'. It could be made clear that any officials and lawyers present are there to be answerable to the public, and that the public, far from being obliged to follow the legal profession's arcane rules of cross-examination, have the right to put their questions and their observations in the common-sense terms that they use in ordinary life. If such proceedings are to be open and democratic in any real sense, it is clear that the object should be to ensure that the public have, *as of right* and without payment, access to the same information and administrative resources as do the paid professional representatives of developers and of the local authority. On the one hand, severe limits would have to be placed on spending on legal representation, whether by developers or by the local authority, while on the other hand public funding of appellants and third parties who can show that they need this would have to be not only mandatory, but generous. The former could pay for the latter.

Such arrangements, I suggest (and one could suggest more), are simply the fundamental preconditions of a public local-inquiry system that promotes the public interest, as opposed to one that promotes the personal financial and professional interests of lawyers, developers and local-authority officers.

That the procedures governing inquiries into such major schemes as road proposals are questionable not only in terms of democracy but also in terms of simple logic is of course well known. That it could ever have been suggested, for example, that matters of national transport policy should be investigated as if they were matters concerning *localities* seems almost unbelievable (see, e.g., Tyme 1978). Yet the system still continues, more than twenty years after this was pointed out.

Another example concerns a statutory right for third parties to appeal against the granting of planning permission. That such a right does not exist provides further evidence of the lack of democracy in the present planning system. Subject to appropriate safeguards, and in defined circumstances as, for example, where planning permission has been granted contrary to the provisions of the development plan, the introduction of this statutory right is among the minimal preconditions of a planning system that could be called open or democratic.

A further example concerns what planners still refer to by the dreadful name of 'public participation exercises'. In a planning system with any claims to democracy it would be illegal for local planning authorities to engage in these, for they are in effect exercises in 'public relations' and manipulation and, being extra-legal, confer no rights on us anyway (Davies 1972, Dennis 1972, Ward 1994; for an opposed view, see, e.g., Crawley 1992).

Instead, democratization of the planning system would imply very considerable strengthening of the relatively few statutory rights that we *do* have, such as the right to inspect and copy documents, the right to have our written representations on planning applications placed before the planning committee, and the right to demand that the planning officer clearly explains to committee why he does or does not consider that those representations should influence the decision made (Reade 1994). The existing statutory right to be present at planning committee meetings could be expanded into a statutory right also to speak, and the right to inspect specific documents expanded into a simple 'freedom of information' right to read and make copies of the entire contents of all files. A legal obligation should also be placed on officials to place a record of all telephone conversations on the files.

When one considers what changes would be necessary if planning were to be made open and democratic, one realizes that the list is almost endless, but it does seem useful here to mention one further matter. In some parts of Britain at least, one finds that ordinary people simply take it for granted, as a matter so obvious that it hardly needs to be said, that Freemasonry plays a larger part than is proper in local government and in planning. Clearly, we do not know whether this assumption is justified, but that it exists is disturbing, and in itself justifies investigation.

There is a good case, of course, for urging openness and democracy as values in themselves, but this is not what I have done here. Instead, I have identified openness and democracy as empirical preconditions of moving from the present token system of planning to an effective one. I would argue that only the public questioning that openness and democracy promote could produce the clarification of aims that is essential to an effective planning system. Only openness and democracy could clarify what planning is for.

3. *To move from the existing token system of planning to an effective one, it will be necessary to abolish the legal fiction that asserts that planning is done for the sake of 'good planning', and that in making planning decisions no account must be taken of their economic and social effects.*

This legal fiction is a nonsense. To show that it is a nonsense, I offer four pieces of evidence. First, common sense tells us it is a nonsense. Any system of environmental planning is a form of governmental intervention in the property market and in the construction industry, both of which are central to the economy as a whole, and to engage in these forms of intervention while supposedly ignoring the economic and social effects of this intervention is plainly irrational. By making clear all the complex and multitudinous ways in which our lives are inextricably bound up with the surroundings in which we live, common sense also shows us that our physical surroundings are of enormous social and economic importance to us. That we pursue our social and economic needs, wants and aspirations through modifying our physical surroundings, and that our lives are significantly shaped by the ways in which *others* do so, is made wonderfully clear, for example, in the radio journalism of Ray Gosling or the published journalism of Colin Ward; in the pages of the *New Statesman* and *Town and Country Planning*, Ward regularly provides us with a continuous stream of evidence, showing beyond doubt that desirable environmental, social and economic initiatives can only sensibly be pursued jointly.

Second, there is the example of the early proponents of town planning. Far from urging that no account be taken of the economic and social effects of the physical patterns they advocated, these physical arrangements were mooted precisely *because* they would have specific economic and social consequences. Ebenezer Howard, for example, whose efforts led directly to the establishment of the Garden Cities Association (now the Town and

Country Planning Association (TCPA)), urged the development of garden cities as a form of dispersal of population because this, by reducing population pressure and thus demand for land in London, would diminish the economic power of the landlord class. As land values were thus progressively transferred from London to the garden cities, he argued, these increases in value could be collected in the form of 'recoupment' by the local authorities of his garden cities, and spent on social provision.

Third, we have the fact that the present British planning system was established by the Attlee government precisely in order to effect social redistribution. It was intended, for example, to increase the real incomes of the less affluent, of which good physical environment is a part, by ensuring that they were enabled to live in desirable physical surroundings. And this legislation rested centrally on the expropriation of development value in land, which is itself a significant form of economic redistribution.

Fourth, the environmental movement that has developed over the last three decades has also found that physical-environmental questions simply cannot be separated from questions of economic and social distribution. Advocates of 'sustainable development', for example, rapidly found that the questions they address are inseparable from questions of social justice (see, e.g., Elkin *et al.* 1991, Blowers 1993a). Similarly, 'greens' have concluded that, of necessity, 'green' implies 'red' (e.g. Pepper 1993).

It would not be possible to move from the present token system of planning to an effective system, therefore, unless we first decided exactly what economic and social objectives this future system of environmental planning was intended to achieve.

4. *It makes no sense to have any physical planning system at all unless it is intended to use planning powers to produce greater social equality.*

Strictly speaking, of course, I should have said '. . . unless it is intended to use planning powers to produce a pattern of distribution of environmental and economic benefits other than that which the market would produce'; intervention could be used to *increase* inequality, as well as to reduce it. Public announcements of government intentions to exacerbate social inequality, however, are not something that polite conventions permit. If one is happy with the pattern of distribution of environmental and social costs and benefits that the market produces, however, then clearly one needs no planning. The only reason for having a 'token' system, such as presently exists, is to conceal the workings and the effects of the market. The existence of this 'token' system enables us to claim that we have 'curtailed the worst excesses of the market', or 'safeguarded the public interest against selfish sectional and private interests', or however we choose to express it, but it does not, unfortunately, enable us to achieve credibility by demonstrating *exactly how and to what extent* we have done so. And this 'token' system has a second disadvantage. Behind the 'smokescreen' that it provides, the opportunities for pursuing private as against public interest may be even *greater* than in a free-market situation with no 'planning'. In a totally unplanned free market, after all, there are no publicly funded institutions of concealment. If, however, there *is* a political commitment to promote greater social equality, it makes sense to have a physical planning system. For the use of physical planning powers provides *one* of the methods whereby greater equality can be effected.

Though my object here is merely to spell out logical and causal relationships rather than to urge either planning or reduction of inequality, it does seem appropriate at this point to mention the sheer scale of environmental inequalities in Britain. These are enormous. Though we have had a supposedly effective planning system since the Second World War, one of its most striking features is its failure to prevent the creation of slums. It would be mistaken to suppose that this Victorian word describes only Victorian phenomena. Today's British slums are clearly not places whose inhabitants go hungry, cold or deprived of fresh air and sunlight. They are, however, places whose inhabitants are profoundly stigmatized and alienated, and who are deprived culturally, educationally, socially, politically and financially; we experience a sense of alienation and deprivation precisely because we are perceived as living in such places. Middle-class people probably find it hard to appreciate quite what it means to be thus deprived of self-esteem by simple association with a *place*. There are areas in Britain, for example, where petty theft and vandalism are endemic, where no amount of parental care can protect young children from exposure to drug dealers, where teenagers grow up so educationally, culturally and environmentally deprived that they get their kicks by stealing and racing their neighbours' cars, and thus sometimes killing and maiming these neighbours and their children – where, in summary, it is impossible, even for those whose personal standards of behaviour are very high, to avoid slipping downwards into the 'underclass'.

And there are quite other places, 'nice' places, where quite other young people, from good homes in nice suburbs, engage in such pursuits as creating nature gardens or repairing footpaths or digging wildlife ponds or planting trees, interacting with and learning from their environment and from other people in ways that help make their lives rich, full and rewarding.

These are worlds apart. They are quite as far apart, I suggest, as they were when Dickens was writing. The forms in which inequality is expressed have of course changed and, in particular, the levels of consumption of both the 'haves' and the 'have-nots' have increased enormously. Inequality itself, however, is as great as ever, and only massive governmental intervention could reduce it.

If it be accepted that it makes little sense to have a planning system at all except as a means of reducing social inequality, then, and if it be decided in the light of this to have a planning system, the next five points (5 to 9 inclusive) follow logically.

5. *A central aim of an effective system of environmental planning must logically be to reduce car-dependence.*

At first glance, this may look like an unwarranted jump from matters of pure principle into matters of great specificity. But it is not. It follows automatically, as I have said, as night follows day. Given that it has been accepted that the only reason for having any physical planning system at all is to effect social redistribution, and *if* it has been agreed, in the light of this, to have a physical planning system, *then*, I suggest, my point (5) becomes a simple truth. Physical planning cannot pursue this goal of greater equality by *any* means it chooses. It cannot pursue it by means of an incomes policy, for example, or through the tax system, or through the education system. It can only pursue it by acting on *environmental* inequalities. To say this, however, is not to say that the contribution of the physical planning system to the reduction of overall inequality must of necessity be modest. It could be very significant, for specifically environmental inequalities in fact often constitute a very large part of overall inequality. We devote a significant part of our disposable incomes, after all, to securing (or as I shall argue, vainly attempting to secure) those things that planning law delightfully sums up as 'amenities'. And we do so to a large extent through use of the car.

The car, I suggest, is probably the chief means by which environmental inequalities are created and sustained. In using their cars to get to and from the relatively tranquil or secluded places in which their homes are set, for example, the privileged impose noise, fumes, danger and all manner of other unpleasantnesses and inconveniences on all those living in the less socially desirable areas which they drive through.

We might characterize the workings of a car-based system of social stratification as a 'game' in which the object is to ensure that all the various activities that make up one's life – work, home, friends, children's school, shops and so on – are in 'nice' places, while one only sees the grotty places through one's windscreen as one moves between these various nice places. But this game must inevitably have far more losers than winners. While some manage only *ever* to see the grotty parts through their windscreens, others, less fortunate, have more frequent and more tactile contact with them, while others again, even less fortunate, must actually *live* or spend the greater part of their lives within these places of stigma.

The strongest argument against our playing this game with each other, however, is not merely that most of us end up as losers. There is an even stronger argument. It is not a static game, but a dynamic one; they keep moving the goalposts! The mere fact of using our cars, to play the 'game', constantly renders *more and more* places suitable only for nice people to drive through and not to live in. For the sad fact is that traffic is probably the chief cause of environmental degradation.

6. *A central aim of an effective environmental planning system must logically be to promote alternative means of transport.*

It is the dominant means of transport in any society that largely determines its whole socio-spatial pattern. If we wish to secure a specific socio-spatial pattern because this reflects agreed social values, as the 1947 planning system did (Hall *et al.* 1973), it seems less than logical to pursue this aim by controlling the socio-spatial pattern itself. It would be more effective to control the means of transport. It is, after all, generally more sensible to act on causes than on consequences.

One of the fundamental objections to the free market is that because it tends so strongly to concentrate economic power, it reduces choice. Where the unregulated free market reigns, as in the United States always and in Britain under Thatcherism, we have little choice other than to accept the spatial separation of our homes and workplaces that the car-based economy produces. Since this is the socio-spatial pattern which maximizes the profits of those interests that monopolize economic and political power, no other pattern is on offer (see, e.g., Hamer 1988). We are thus forced, as a direct result of living in a society in which those who control the vehicle and roads industries have so much power, to devote a relatively large proportion of our incomes to the purchase and maintenance of cars and roads.

To live an alternative lifestyle, based on public transport and the close proximity of all the various activities that make up our lives, is clearly impossible where, due to the power of the roads lobby, these things are *not* in close mutual proximity. Quite simply, the very considerable economic and political power of this lobby make it increasingly difficult for us to choose any lifestyle which is not centred on consumption of its products.

One of the main reasons for having environmental planning, then, is to counteract this monopoly of economic power, and to restore choice. Promotion of public transport, and the creation of attractive 'green ways' enabling us if we choose to get around on foot or on bicycle without having to suffer the unpleasantnesses inflicted by motor vehicles, must therefore be among the central aims of the planning system (see, e.g., Hass-Klaus 1990, Hillman *et al.* 1990, Tolley 1990).

I wish to emphasize here that 'theses' 5 to 9, which all centre on reduction of car-dependence, follow simply from the prior decision (provided it were taken) to adopt planning in order to reduce social inequality. There are of course *other* arguments for reducing car-dependence – it causes pollution, is a major cause of death and injury in society, and makes it virtually impossible for children to explore their surroundings other than as passengers in their parents' cars – but I am not concerned with these other arguments here.

7. *A central aim of an effective system of environmental planning must logically be to challenge the assumption that we must inevitably and unceasingly be seeking a 'nicer' place to live, in order to 'better ourselves'.*

This widespread social assumption too is a consequence of motorization, and thus of the dominant position in the power structure of those who control the vehicle and roads industries. The dominant ideology in Britain, however, portrays it as a law of nature – or of human nature. It asserts that the process whereby those with any ability are constantly moving on – and thus leaving the less adequate increasingly concentrated in the 'inner cities' or other places defined and perceived as socially undesirable – is 'natural' and inevitable.

It is no such thing. It is simply a reflection of the power of those who control the vehicle and roads industries to persuade us that our self-esteem depends on our consuming the things they sell us. The most important of all the things they sell us is ironically not cars, or even good roads, but 'nice places'; possession of a car is essentially a *promise of access* to places socially defined as desirable and enviable, whether for commuting from and thus 'living in', 'escaping to' at weekends, being a frequent visitor to, or whatever. This 'move-on' culture is economically wasteful and unsustainable. The lifespan of the physical fabric of the areas that the 'successful' have moved out of or avoid is unnecessarily shortened, since it is difficult to attract investment into places that have been rendered into social symbols of personal failure.

8. *A central aim of an effective system of environmental planning must logically be to ensure that all can live and work in pleasant surroundings, and without moving house.*

This follows from the previous point. Surroundings socially perceived as desirable (*not* 'good surroundings') should be seen as a fundamental right of all citizens, not as something that we must constantly pursue with the help of the internal combustion engine. This aim should be achieved not by helping us to move to 'better' surroundings, but by using planning powers to ensure that *all* places are pleasant – or at least that no places are so unliveable that we feel we have to move out of them for that reason alone. Good environment, in other words, should come to us. We should not have to chase it. The pursuit is in any case self-defeating, as I have shown under thesis 5; in constantly using our cars in order to be in 'nice places', we destroy the niceness of all places.

This suggests we need a new basis for the use of planning powers. At present, the planning system seeks to provide us with a good environment by controlling development. Environmental deterioration, however, is something that occurs quite independently of development, and affects far wider areas than the places where development occurs. And it is environmental deterioration rather than development that is often the real cause of decline in the quality of our lives. Indeed, we might reasonably say that development that threatens to reduce the quality of our surroundings is best seen as merely *one particular kind* of environmental deterioration.

Much of the development that the present planning system controls is in any case relatively trivial, and it is arguable whether this relatively expensive existing form of control makes our surroundings significantly better than they would have been in its absence. Planning could instead rest on a statutory public right to demand a public inquiry wherever *any* factors, including proposed development, seem to threaten the quality of our local environment. Clearly, such a legal right would have to be carefully defined, and wherever possible expressed in quantitative terms. But it could, for example, be made to depend on our demonstrating that traffic noise had increased to a defined level, that the percentage of empty properties had reached a certain figure, or that our neighbourhood suffered from a specific level of vandalism, physical decay, fly-tipping or littering. Such a right to require a public inquiry, chaired, for example, by an independent planning inspector as are inquiries into planning appeals, would seem likely to provide a better guarantee of good environment than does the present system of development control. It would also be more likely to encourage public involvement in the shaping of our physical surroundings, and more generally to encourage in us a sense of social responsibility and civic-mindedness. It would, after all, make us in a very real sense – in a legal sense – the guardians of our surroundings.

To base planning on this statutory right to demand intervention would probably also reduce the *need* for such intervention, since local authorities would soon find timely vigilance and good environmental management cheaper than public inquiries leading to relatively large-scale remedial measures; prevention is usually cheaper than cure. But where relatively large-scale environmental investment *was* required, it could be afforded; taxation of land value, which is the most logical way to pay for public environmental improvement, would provide our 'effective' planning system with a very considerable income (Reade 1987).

9. *A central aim of an effective system of environmental planning must logically be to encourage a 'European' as opposed to an 'American' lifestyle.*

This may seem both polemical and provocative. It is not intended to be. These two terms, I suggest, are simply appropriate names for two identifiable and contrasted characteristic socio-spatial patterns, which empirically exist. And this ninth 'thesis' is in any case no more than a summary of the previous points (5 to 8). In any society in which the vehicle and roads industries are as economically and politically powerful as they are in North America, the socio-spatial pattern which I term 'American' results automatically if we simply allow the market free play. We therefore need no planning system to achieve it. We only need a planning system if we wish to counteract or reduce the prevalence of this pattern.

The 'American' model is characterized by almost complete car-dependence, a consequently highly dispersed settlement pattern, almost complete reliance on a largely unregulated free market, and relatively low levels both of taxation and of social provision. The European model, by contrast, is characterized by more or less strong measures to discourage car-dependence and thus to preserve the urbanity and liveability of historic towns and cities, relatively concentrated settlement patterns, a type of capitalist organization that is usually to a greater or lesser degree corporatist, and relatively high levels of taxation and social provision. The American model puts high value on personal freedom and gives little weight to the need for social redistribution, while the European model uses the tax system and other mechanisms to promote social equality through economic redistribution. These two models are of course no more than ideal types, against which actual situations can be measured. But each of them has sufficient coherence and internal consistency to warrant our identification of them as alternative models; the internal components of each model tend empirically to imply each other.

The value of these two models, I suggest, lies in their reminding us that all environmental choices must inevitably be rooted in values. They also help by showing us the need to research empirical questions. Could we, for example, choose to have bits of the one model, and bits of the other? I would hypothesize that in the main, we could not. What is of particular interest about these two models in the light of the arguments advanced in this chapter is that they relate to two cultures with rather similar levels of wealth and economic development. That these two cultures can be characterized by such very different socio-spatial arrangements, therefore, and that their citizens use their wealth in such very different ways in accordance with very different value systems, suggests that political choice is, after all, possible. We *can*, it would seem, and despite what Thatcherism tells us, collectively decide what kind of culture we want.

Thatcherist ideology, of course, denies this. It tells us that 'there is no alternative'. It uses false determinism to persuade us that the 'American' model reflects 'human nature' rather than a particular configuration of power. As regards what I have termed the 'European model', it seeks to persuade us that this is not a pattern that rational beings would choose. It portrays it as an unnatural configuration in which political interferences produce economic distortions.

SUMMARY AND CONCLUSIONS

If Thatcherism has taught us anything, it is that ideology is more important than we may have supposed. Ideology, it would seem, is not necessarily something constructed *ex-post facto*, to legitimate and justify power already achieved and exercised. On the contrary, it can *achieve* power for those who deploy it effectively, or at least it can be used by those who already have some significant degree of power to vastly increase that power. It is this that is meant when ideology is defined as 'the power to create facts', and it was with this in mind that I described the social transformations effected by Thatcherism as 'propaganda-led'.

Yet strangely, this lesson concerning the potency of ideology is the one lesson *not* generally drawn from Thatcherism, at least by the Labour Party or at least not yet. Though of course its tactics may change, the Labour Party has so far not yet given much sign that it seeks to achieve power and to effect social reform by propagating an 'alternative ideology'. Instead, it appears to be seeking power by portraying itself as committed to values not strikingly dissimilar from those of Thatcherism itself.

Since their project rests centrally and unavoidably on the dissemination of an alternative ideology, this creates problems for those who wish to see an effective system of environmental planning. *No* political party in Britain evinces much interest in creating a future planning system legitimated by being explicitly based on values summed up by the concept of 'sustainability', or on 'environmentalist' or 'green' ideas, and *no* political party offers either to place the physical planning system within a wider system of economic and social planning, or to provide the planning system with an economic rationale by bringing back effective mechanisms for the recoupment of development value.

Virtually the *only* ploy available to those who wish to see an effective system of environmental planning, therefore, lies in their allying themselves with those campaigning groups which, unlike the Labour Party, *do* offer a real alternative to Thatcherism. This means their joining forces not only with all those groups which campaign for environment in the widest sense, but also with those who campaign for constitutional and electoral reform, for freedom of information and openness in government, for social banking, Third-World aid and Local Exchange and Trading Schemes, for more rational transport and energy policies, for 'alternative economics', for innovative social ventures of all kinds and, above all, for recoupment of development value in land. The extent to which all these varied branches of the 'alternative culture' share common values is perhaps greater than many have yet realized. The suggestion that the advocates of environmental planning work with them has recently been made by the Town and Country Planning Association (Cordy 1996, Porritt 1996). Though the Royal Town Planning Institute, at least in its published pronouncements, now itself promotes many of the ideas of the above-mentioned campaigning groups and causes, and though the Institute increasingly adopts the attitudes and the methods of a body urging social reform rather than those of a narrow professionalism, it would still seem to be the case that arguments for planning are in the main better promoted by campaigning groups such as those mentioned above rather than by a planning 'profession' (cf. Reade 1987, Evans 1993).

It is just possible, therefore, that an effective 'alternative ideology' might emerge out of such co-operation, that this alternative ideology might begin to supplant Thatcherism as Thatcherism supplanted the 'post-war consensus', and that a changed political climate in Britain might thus emerge. Only if there occurs such a change in the national political and social climate, I suggest, will there be a possibility of achieving real environmental planning in Britain.

6

CAN TOWN PLANNING BE FOR PEOPLE RATHER THAN PROPERTY?

Bob Colenutt

The town planning system has always been divided between its concern for people and the imperatives of the property market. More often than not it seems to be the servant of property; though just sometimes it defends and enhances the lives of ordinary people as well.

In the post-war period, planning, as a policy instrument, was widely accepted as part of the welfare state. It was a tool for social as well as spatial distribution. Land use, and land values, are created in part by the wider comunity so landowners had to operate within a socially defined planning framework where their needs were only one part of the equation.

Landowners did not accept this social contract without protest and have sought to shift the balance of power towards the market place. However, the balance of forces in planning has now reached the point where town planning is firmly locked into the demands of property and landowners so that it rarely seems to stand for anything else but facilitating property-market demand. This is not because developers ignore plans or circumvent them. This has happened on occasions, but the changes are more fundemental than that. The government has cleverly altered the policy content of plans so that, although development is now 'plan led', the fact is that most development plans reflect central government social and economic policies rather than the wishes of local communities.

This chapter explores the balance of power between the property market and the social and environmental aims of planning, and in particular examines the taboo about planning as a force not only for physical changes in the environment but for social redistribution as well.

THE PLANNING SYSTEM

Planning has been on the retreat over the past twenty years and is possibly in a weaker position than at any time since the creation of the post-war town and country planning legislation.

There are three factors that underlie the progressive weakening of the UK planning system. The first arises from the fact that town planning is a profoundly political process that reflects the political and economic balance of power of the day. Some planning professionals have argued that planning is neutral and should be bipartisan, accepted by all political parties. Of course, it is not. The Conservatives have been vigorous in their attacks on planning since 1979 as Thornley (1991) and others have shown. Deregulation has sharply reduced public (and locally elected) control over development (far more decisions for example are now delegated to officers). Local authorities have fewer resources to take on the spiralling costs of fighting planning appeals, Urban Development Corporations (UDCs) have taken away local authority planning powers in fourteen key regeneration areas in England and Wales, and over thirty enterprise zones were set up which for ten years removed planning control altogether. As in many other areas of public life, government officials in regional offices, quango board members, auditors and government inspectors wield far greater power than ever before.

Planning is, therefore, very far from neutral or purely technical. This goes for the small print as well. A close reading of a Unitary Development Plan or Strategic Guidance Note will show just how much the political ideology of the Right (e.g. the effective presumption in favour of the developer) is now embedded in the planning system.

The second reason for the weakness of planning is that it represents a value system that places markets above people. There is an implicit social, economic and environmental vision behind any planning policy document. The value system that is expressed in most plans or at planning inquiries largely accepts market objectives and the physical manifestation of development. Non-market values centred on people and social or cultural needs play only a minor part in determining planning decisions.

A third reason for the weakness of the planning system is that because it has been oriented towards the achievement of property-market objectives, it means very little to disadvantaged communities, and does nothing to help them with their day-to-day needs. In many parts of the country, development plans and planning decisions, if anything, seem to make the conditions of life worse.

At one time, this judgement would have been made about local-authority plans for 1960s tower blocks and faceless shopping centres. Now this view applies to 1990s city centres, to new urban settlements in the countryside and to the Isle of Dogs, where large-scale development has run riot, local people have little or no control over what has happened, and there are precious few community benefits.

PLANNING AND COMMUNITIES

My experience of planning is London based, where competition for land and property is intense, property values are high, and where planning decisions can create huge increases in land values without a brick being laid. In these circumstances, planning can make a significant difference to everyday lives. Yet community development and economic change in most neighbourhoods is rarely seen by local residents as being created by town planning. It is more likely to be seen as caused by more tangible factors such as job losses, housing shortages, transport costs, social and cultural changes, local politics and so on. Very rarely is there any awareness that changes or threats to the environment or the structure of the community are created by planning, or can be resolved by planning.

However, there are exceptions. Along London's South Bank and in Docklands, for example, community action sprang up specifically around planning and redevelopment issues. The booms and slumps of the property cycle have been intensely felt in these areas for well over twenty-five years. They are located in the so-called City 'fringe' areas where developers want to build when the market is up but won't touch with a barge pole when the market slumps. If you live in these areas you know all about the ups and downs of the commercial property market. You can see it in the cranes and the 'To Let' signs.

The property booms of the 1970s and 1980s transformed working-class communities like these into office and luxury-housing zones. When the slump came there was massive oversupply of office space. But amid the confusion it was clear to community leaders that a lot of power was vested in the planning system. It appeared to determine the value of land, the cost of social housing, and the survival of small firms, parades of shops and community halls. For the developers, town planning was a means to an end – a series of hoops to pass through in order to make a profit. For the community, getting involved in the politics of planning was the only way to begin to claw back land and buildings and space for the community. Thus, to win political control and influence over the local and regional planning system was a key goal of community action.

When I began working for the North Southwark Community Development Group (NSCDG) in London in 1972 in the teeth of the last but one property boom, we were not fully aware of all this. North Southwark is a hard-pressed inner-city community with very serious problems of housing and social distress. The planning system had a huge impact because individual land-use decisions made by the officers and councillors in the town hall had direct implications for people's lives, for their schools, shops and employment.

Without a planning background, neither I nor my colleagues at the

NSCDG were aware of the significance of the planning system. We barely knew what a planning application was, let alone a planning appeal, a statutory plan or a stop notice. We didn't know about development plans, the Greater London Development Plan or the previous development plans. These were all intimidating mysteries but we realised that we had to find out if the community was to fight back. And we did find out, and began a long struggle aiming to change the local and London-wide planning system so that it defended the interests of the community rather than the interests of the land and property market.

Our main discovery was that behind the bland planning documents and their dry texts was something far more important than planning itself. Planning was very closely and intimately tied up with the workings of the land and property market. In fact, it was a product of the property-market system. Its function was to facilitate development, to make it happen on the ground, and to give it spatial configuration.

Even more than that, planning gave a legitimacy to market forces and at the same time hid the property market from the people. It concentrated on planning reports and planning applications instead of the calculations of developers about land values, rents and profits. At the planning applications sub-committee, local people and councillors are faced with planning officers and the agents for developers, but not the actual developers or landowners themselves. The people behind commercial property – insurance companies, banks and pension funds, unit trusts, development companies – are largely hidden. The planning system thus protects private landownership and the financial web that is connected to it from exposure to public scrutiny.

Rarely in North Southwark, or along the South Bank, or in Docklands did the landowners or developers lose their arguments with the local planning authorities. They almost always got their way, or if they didn't, they came back again, and they got their way next time around.

The lobbying power behind land and property markets is enormous. Not so much the direct lobby agencies for property such as the British Property Federation and the Royal Institution of Chartered Surveyors; they use their influence as you might expect. But there is a prevailing ideology supporting property owners and market forces which has moved very quickly to resist any attempts to curb the land and property market. Private landownership and commercial property have enormous credibility in the UK.

Thus, to change the planning system to the needs of communities, or to a more locally accountable basis, or to tinker with the Town and Country Planning Acts, or to issue Planning Policy Guidance that goes against the grain of the market, means you are up against heavyweight market and professional interests; not to mention the media. It is a mistake to think the planning system can be reformed simply by lobbying associations of professional planners or civil servants. Planning is a highly political issue with huge amounts of money and vested interests at stake.

COMMUNITY STRUGGLES

Local communities like those in North Southwark or Docklands have no option but to try and gain democratic control over the planning system if they want to achieve community and environmental goals. Even though control over neighbourhoods and their environments goes far beyond the planning system, local planning can be an important starting point. Indeed, local authorities will often claim that they are using their planning powers, through development plans and control of planning applications, on behalf of local communities. But it is unwise to accept this claim at face value.

Local authorities of necessity have a corporate (local-authority wide) agenda. Communities must themselves engage in the politics of planning at both a local and national level if they want to resist commercial development schemes or undertake community development that goes against the grain of the market. They will need the support of local authorities, but they should not leave the arguments or the politics to the local council. The participation of local residents makes a difference to how the local authority conducts itself and can also affect how the developers and landowners act. Local residents are the unpredictable element. Developers and local authorities are familiar with each other. Local communities will not compromise so easily and will keep on asking the difficult questions. Moreover, local communities are far more likely to have the energy and imagination to come up with alternative proposals.

In some important cases such community action has achieved remarkable success. For example, in London, at Coin Street on the South Bank, community groups aided by the GLC and the local authority were able to halt a major office development in the early 1980s. The GLC bought the 13-acre site and handed it over to a community trust. This trust has already built over seventy low-rent houses, a public park and riverside walk and has refurbished (for flats and workshops) the landmark OXO tower building, a former coldstore on a prominent riverside site close by the National Theatre.

None of this has been easy. It has been a long battle, funds have been short, the banks have been unwilling to lend to newcomers without a track record. Even more difficult have been the compromises required to obtain grant funding for housing and refurbishment from Conservative government ministers. There is the ever-present danger that in the determination to achieve development on the ground, the community development trust will turn into a quasi-private developer and lose its base in the community struggle for social housing that brought the Coin Street Action Group about.

Similarly, at the former Courage Brewery site in Southwark opposite St Paul's, a community campaign calling for 'homes not offices' persuaded the GLC to buy a key riverside office site from the owners during the property recession of the early 1980s. The GLC then passed it onto Southwark

Council who built houses with gardens for rent across the site.

In both these examples, the key ingredients were highly motivated local campaigns and, at the same time, close co-operation between communities and local authorities. They skilfully used the planning system to oppose speculative offices and pave the way for social housing, community centres and local workshops. The collapse of property values in the early 1980s recession was crucial as well, but the planning battles between local authorities and local residents over development plans, and individual planning applications, eventually involving long-drawn-out public inquiries, held up market forces long enough and laid the planning foundations for community-led schemes to go ahead.

Similar planning campaigns have been fought elsewhere in inner London, for example at Battersea Power Station, the King's Cross Railway lands, Fitzrovia, Hammersmith and at Covent Garden. The planning arguments, proposals and counterproposals can go on for years, through each boom and bust cycle with neither market nor community forces able to clinch a clear victory.

Such campaigns are unusual. Land-use conflicts in other parts of London and in most parts of the country are not so polarised. Many campaigns are short lived. Yet those that do manage to survive beyond initial protest create political awareness. The key role of planning committees, elected councillors, planning officers, inquiry inspectors, consultants, landowners and developers becomes clear, as does the power exercised through Government Planning Guidance Notes and ministerial decisions on precedent-setting planning appeals. By creating a high political profile on local planning problems, the community can air wider issues that are not often regarded as 'material planning considerations', such as homelessness, unemployment and community decline, and the quality of public transport.

The goal for communities engaged in conflicts over land is not just to win the technical argument (important though this is) but to build community strength and confidence. In other words, even though the planning system is not designed to protect inner-city communities, it is possible through intensive long-term campaigning to win something and to use this as a basis for wider community development.

The fact that community victories against the market and other powerful forces are few and far between demonstrates that the balance of power in the planning system is clearly tilted to the market/government, although the balance swings with shifts in political forces and with the boom and slump of the market. During the late 1970s and 1980s in London it swung to some degree towards the community, as shown by the relative success of the GLC's Community Areas policy and campaigns like that at Coin Street.

The Community Areas policy aimed to restrict office development in neighbourhoods facing office-development pressure, while at the same time providing funds for housing and commmunity projects in those areas as an alternative to market-led development. The GLC used its considerable resources to buy out key sites such as Coin Street, the Courage Brewery and Camley Street at King's Cross to give community plans a chance of being implemented. Community development in the Community Areas meant much more than just saying no; it enabled people's plans to become a reality and demonstrated that, if backed by funding and political support, they have enormous potential.

PLANNING AND SOCIAL CLASS

An enduring characteristic of planning is its class bias. This subject is not an easy one to discuss since many people immediately demand to know how class is defined, or argue that there is no longer any such thing as class, or say that other social and cultural distinctions are more important. Certainly class is difficult to define; it has changed, and other distinctions are very important as well.

Nevertheless, there is a social-class bias in the planning process. The strong professional and legal composition ensures this. Just go into a planning inquiry and you know it's there, from the lawyers downwards; look around a run-down housing estate; map where the conservation areas are and who lives there. Map where the planners themselves live. Map where the owners of investment in property live. Power and wealth in the land and property system, and hence in the planning system, are distributed roughly according to class. It is not a perfect correlation, and there are many local differences, but looked at overall, social class calls the shots.

Not only are the structure and assumptions of the planning system biased against poorer communities (and conversely in favour of the better off), but there is a double standard in the practice of planning. Similar proposals are often treated in different ways in wealthier compared with poorer areas. Building projects, roads or redevelopment proposals that would (more often than not) be thrown out by the planners, politicians and Department of the Environment (DOE) Inspectors in, say, Surrey or Hampstead or Bromley, are given the go-ahead in the inner cities or peripheral estates, in spite of local protests.

How is it that such double standards exist? It is sometimes alleged that inner-city communities value their environment less than those from leafier suburban areas, or that they have less to lose. On the contrary, they have more to lose because they start off with a poorer environment in the first place. Yet the government listened more to the people of Kent fending off the Channel Tunnel Rail Link than, for example, to the people of Docklands fighting against the Docklands highway, which carved through the middle of Poplar in Tower Hamlets, demolishing 500 homes in its wake. Planning decisions thus often reflect local voting patterns and social prejudices rather than the merits of developments.

This class bias is also linked to the definition of what constitutes a 'material planning consideration'. Many of the issues that matter most to inner-city communities, for example whether a new scheme is for affordable homes or for luxury flats, or whether an office block is going to create local jobs, or whether a new hospital is private or public, are not considered under the planning laws to be material in judging a planning proposal, because they are not strictly 'land-use matters', i.e. matters relating to the physical environment. It gets worse because key issues of who owns buildings or land, or who benefits from the development, or whether it is viable are also ruled out. Thus, many of the social and environmental problems that matter most to local people (and the questions that most readily come to mind about development proposals), do not form the basis of assessing a development when it is considered by the planning authority or planning inspectors from the DOE.

Planning is, therefore, in many ways simply not relevant to many people. It may be relevant to those living in middle-class areas protecting their property values and their environment, but it does not often seem a useful way of defending the environments of poorer communities in the inner cities or peripheral estates. If anything, planning is part of the problem. A geographer, William Bunge, uses the expression 'if it's not useful it's useless', and this is what many feel about the planning system. It is certainly not useful, and in many cases it is completely useless for poorer communities.

The conclusion is that we need to create a planning system that *is* relevant to those communities. Planning should be turned on its head into a tool for positively improving the environments and opportunities of the residents of disadvantaged neighbourhoods.

CONFUSED SIGNALS

Some observers say that after the 1980s 'nightmare years' for planning, we are at a watershed; that the experiment in deregulation and anti-planning is now over. Some argue that planning is making a comeback. Even middle-class neighbourhoods have begun to complain about deregulation, urban sprawl and motorway building. Most interesting of all, the development industry itself is calling for planning to give it some certainty about future land uses and provide a spatial structure for development proposals.

Recent debates about sustainability and Agenda 21 have suggested that town planning could potentially become the key policy agent for promoting sustainable development. In theory, this policy shift could be a significant challenge to market forces and ultimately to government policy on key topics such as transport, energy policy and land-use allocations.

A plethora of planning documents has been published on sustainable development by national and local government, by the European Union and environmental pressure groups, and there has been speculation that this could lead to a clamp-down on out-of-town shopping centres and car-based residential and business development. Yet in spite of this, and the accompanying tide of public concern for the environment, there has yet to be any significant change in the way the property market operates in practice. Its narrow investment criteria, preference for 'out of town' as opposed to 'town centre', and unwillingness to take risks in more marginal areas are the dominant features of the market.

From a local authority perspective, although Councils *want* to be more green, it is very difficult to achieve this in practice. No local authority operates 'deep green' planning policies, and if one did, it is unlikely that it would be upheld at planning appeals or public inquiries. Furthermore, just at the time that the green debate is taking place, 'regeneration' has come along as another urban-policy objective which, instead of reinforcing green objectives, is actually undermining them – for practical economic reasons.

Local authorities faced with high levels of unemployment and low incomes are desperate to attract development that brings in jobs and private investment. For example, a proposal for a hypermarket or a 'leisure box' with a multiplex cinema on a derelict out-of-town site may well be permitted as a departure from sustainable policy in spite of the fact that it includes 500 to 1,000 car parking spaces and lies outside the town centre. Faced with the possibility of 200–300 jobs and the removal of a depressing, bad-for-the-image eyesore, the temptation to make the superstore an exception to the Local Plan is overwhelming.

Many observers had hoped that this type of demand-led planning would end with the phasing out of urban development corporations (UDCs) and enterprise zones. They looked to the introduction of City Challenge and the Single Regeneration Budget (SRB) in the early 1990s as indicative of a new emphasis on planning in partnership with the local community. Certainly, the SRB broke with the regeneration policies of the 1980s by returning more power to local authorities and integrating under one budget a variety of urban programmes. With a combined budget of £1.3 billion per annum, it offered grants to regeneration partnerships across the country and though allocation of funds was on a competitive basis, most local authorities and community organisations embraced SRB with enthusiasm. Its purpose, to integrate physical development with training, community safety and housing programmes focusing on a defined local area was welcomed, as was the requirement to involve the voluntary and community sectors.

Yet the SRB (renamed the Challenge Fund) remains at heart a property-led programme which aims to lever in private-sector investment – and this means allowing the market to determine the type and pattern of regeneration. Relatively little SRB is directed at ensuring that there are tangible benefits to poor and disadvantaged communities arising from property investment. It is true that linkage between physical schemes and social deprivation is somewhat better than UDC-style development but, in practice, local employment benefits for disadvantaged communities are very limited.

Moreover, the amount of funding going directly to community-led schemes is little more than 10 per cent of the total programme.

Thus, the experience so far with sustainable development and regeneration suggests that although the crude market-led planning of the 1980s is over, the planning system of the 1990s is also market-led (but more subtly so). Though tacitly acknowledging the social and environmental implications of physical development, the current policy framework draws the line at social planning and does not seriously grapple with social deprivation and disadvantage and the empowerment of deprived communities. Urban regeneration, with its emphasis on partnerships and integrated programmes, has muddied the waters about the distributive impact of urban development, but it has not changed the central property basis of urban policy.

Moreover, the signals are further confused by a proliferation of community consultations, 'Planning for Real' and Urban Design Action events which bring a welcome grass-roots voice to localities. Schemes like these involve detailed scrutiny by local residents of the strengths and weaknesses of their local environments. They contrast sharply with the content of many unitary development plans that limit themselves to the usual range of planning issues and 'strictly planning' policies. Statutory plans often seem more like 'planning for unreality' than an exercise in engaging with communities on issues that really matter.

This suggests that planning is operating now at two levels. At the informal, very local, level there is evidence of a resurgence of community approaches to planning, which are often explicitly about social and economic redistribution. At a formal statutory level, however, traditional planning assumptions and practices continue, reflecting the interests of the worlds of land, property and central government.

The same twin-tracking can be seen in the debates about 'community' and 'partnership'. Again there is welcome change in rhetoric from confrontation to partnership and community. But power within partnerships for regeneration and urban development is rarely equal. The main players are the development industry, government agencies and local authorities. Local communities, particularly poorer and less organised ones, do not have the resources to participate effectively.

Thus the reality is that power does not lie with communities. And this reveals the real nature of the twin-tracking that we see. Real power lies at the strategic levels, overshadowing and overlooking a small amount of discretionary power at the local community level. As for control over the operations of the market in land and property, local people can do very little. Unless there are exceptional local campaigns, local residents offer little threat to the operation of the property market.

The split between strategic power and local powerlessness is even more sharply defined in London for many reasons. First, this is where property-market forces are most powerful, and where there is most money tied up in property investments. Another reason is that local government in London has been especially disempowered by Whitehall and by the abolition of the GLC in 1986. This has led to fragmentation of services (including planning) and to the creation of a whole strata of government agencies and quangos which act as an unelected government for the capital. Significantly, these quangos have very strong links with the commercial property-development industry.

Thus, in London and elsewhere, planning is back on the agenda, but only in a very restricted sense. Strategic planning concerns itself primarily with city-wide promotion and flagship development. It sees its main task as the attraction of investment for prestige projects whether it be airports, Olympic Games, millennium bids or light rail schemes. Community-based planning exists in a sort of half world beside all this – acknowledged but not properly resourced, and prevented from threatening strategic investment.

PEOPLE-BASED PLANNING

We next turn to the question of what a people-led, community-development-based planning system would be like. Fortunately, we do not have to start from scratch. Many of the elements of people-based planning already exist in countless community projects and campaigns up and down the country. They embody principles and practices that are the starting point for turning planning around.

Second, community planning does not mean the denial or negation of strategic planning nor of wider-than-local analysis. For community planning to be meaningful it must be placed within a context of strategic provision by central and local government. Government must provide essential social and economic services and infrastructure, and take a strategic overview. But this *must* be done in consultation, working with localities. Local residents and workers can and do think strategically as well as locally. There is a myth abroad that only experts and professionals are capable of this.

The purpose of planning, its values and vision should, therefore, be redefined. Communities and their needs should be at the centre not simply responding to the demand (or lack of it) of the property market. If we move down this path, it then becomes possible within the framework of planning consultation to debate explicitly how to create and protect jobs, house the homeless, create a decent healthy environment, ensure adequate public transport and reduce crime. These issues are real, and, if they are not brought into the planning system, town planning will die as an instrument of social policy, leaving it to be manipulated by rich and powerful corporate elites.

We must be able to ask of all planning schemes – who benefits, and who loses? Planning must recognise conflict and construct a balance sheet of winners and losers. It can no longer continue to hide hard choices behind the rhetoric of partnership, regeneration, 'community' or 'City Pride'.

At the same time, for planning to make a bigger and more beneficial impact on behalf of local communities, the land question must be resurrected. Landownership is a major debate in many countries but not in the UK (except to a degree in Scotland) owing to fears of 'land nationalisation'. As Massey and Catalano (1978) have pointed out there is an exceptionally deeply entrenched landed gentry in the UK, to which must now be added the rapidly growing holdings of financial institutions and the new quangos, UDCs, health trusts, privatised utilities, etc.

For communities to develop themselves, they need land and access to land, as well as control over the use of land. There is currently no public policy from government or opposition on landownership except to leave it well alone (unless it is needed for motorways or the Channel Tunnel Rail Link or other strategic schemes). A people-based planning system must have access to land. It is an essential social and community resource.

People-based planning must also have access to funds. Many community initiatives are foundering through lack of public or private capital. While huge shopping centres and giant office complexes and flagship Lottery projects attract billions of pounds in public subsidy, the social and community sectors are starved of resources. SRB and English Partnership funds for community schemes are tiny – a drop in the ocean. The private sector, through pension funds, investment trusts, banks and insurance companies has literally billions of pounds to invest *daily*, yet many worthwhile community social and job-creation schemes are struggling. If it were possible to legislate to divert one tiny fraction of these funds into national or regional community development chests, then many community schemes could become viable.

Planning must also tackle head-on the problems of implementation. A whole range of new grass-roots developer organisations (as unlike the UDCs as possible) is required, including community development trusts and do-it-yourself regeneration agencies. Community development requires resourcing, skills training and access to public and private funds.

An urgent priority is to break the centralising grip of Whitehall, government agencies such as Training and Enterprise Councils (TECs) and highways agencies, and also the often stifling role of local government. Community empowerment is quite different from empowering local authorities, or local hospital trusts, or regional development agencies. Empowering communities means empowering residents acting together. Genuine community and local empowerment is essential for a democratic planning system – not through stage-managed government partnerships such as the Challenge Fund but through giving local residents resources and discretion to act for themselves (sometimes with the state, sometimes through development trusts or public/private-sector partnerships).

Last but not least, we need a new democracy. People's plans are a beginning. They should become as important as local-authority plans: local-authority plans have become too important, yet too remote. Meaningful plans would be those drawn up by local people which include all relevant local issues. Local-authority plan-making is only one form of planning. It may be excellent in a few areas but in most it is not.

How can any of these changes be achieved? Only by campaigning, by highlighting and creating good practice, and roundly attacking bad practice, and by learning the lessons of the past twenty years.

THE LESSONS OF DOCKLANDS

Docklands is a classic case where local authority (and community) power over development has been taken over by parliamentary statute and placed in the hands of the unelected London Docklands Development Corporation (LDDC). The LDDC, established in 1981 and due to continue until 1998, is required to act, not in the interest of local communities but 'in the national interest'. In effect, this means following the market – wherever it may go, whether it is Canary Wharf one year or the recession the next.

The received wisdom about the lessons of Docklands is limited to a general acknowledgement that the transport infrastructure was planned far too late. However, the central lesson was surely that regeneration ignored the needs of local people and failed to involve local residents in decisions affecting their lives and their environment. Second, the deregulation of planning and financial services led to an oversupply of offices because of competition between Docklands and the City of London. The final lesson of Docklands was that the concept of wealth creation (and trickle-down economics) that was imposed on Docklands communities was fundamentally flawed.

The irony was that transport infrastructure was provided in Docklands more quickly and at more cost to the public purse than in almost any other regeneration area in Europe. The LDDC put £850 million into new roads, the Docklands Light Railway was extended and upgraded, and £1.9 billion was invested in the extension of the Jubilee Line to Canary Wharf. Meanwhile, the Department of Transport invested £1 billion in roads leading into Docklands.

The fundamental flaw and ultimate tragedy was the refusal to spend this money on the one million people who live in the five East London boroughs around Docklands. While communities suffered bad housing and deepening poverty, subsidies were poured into luxury development on nearby riverside sites. The community's own plans – for the Royal Docks and Wapping for example – were ignored.

We are in a period of uncertain economic growth. We have reached the end of the domination of monetarism and 'free-market' ideology. But in its place is a policy vacuum, with confused signals.

It is essential that those who believe in the need for generating a new planning ideal, with new values, act quickly. Otherwise, as the market picks up speed, landowners, financial institutions and developers – working with the new city-wide strategic partnerships – will fill the vacuum, and communities will be sidelined again.

7

THE VIEW FROM LONDON CENTRE

Twenty-five years of planning at the DOE

Peter Hall

The year 1995 marked the twenty-fifth birthday of Britain's Department of the Environment (DOE), set up in 1970 as an integrated super-department for all matters of land development and construction, including planning and (for the first seven years of the new department's life) transport. A conference, called by the DOE to celebrate the anniversary, gave a rare opportunity to review the history of British planning over that quarter century, and to speculate on the future significance of the experience.

First, though, a basic problem of sources. Any historian of government finds a rather unusual difficulty: short of privileged access into the files, such as Cullingworth enjoyed in writing his monumental official history of planning (Cullingworth 1975, 1979, 1980), there is surprisingly little basic documentation. The most complete is provided by the DOE's own press releases; and, such is the caprice of fate, the first seven years of these records seem to have disappeared, both from the DOE and from the British Library. So in one important respect, this chapter is fuller for the Thatcher and Major years than for the Heath, Wilson and Callaghan ones. As to the effect of this uneven treatment on the overall judgement, readers are free to gauge.

Skimming through the releases, which latterly have emerged at the rate of 700 a year, does confirm one basic point, well known to anyone within the Department: planning as such has occupied a relatively small space within the total framework of policy making and policy revision. The main lines of the system were laid down in 1947 and have been modified only subtly since then, sometimes – as in the controversial financial provisions – subject to regular political part-song resembling the old nursery rhyme: Labour put the betterment on, Tory take it off again. The main policy initiatives have occurred in other, often closely related fields: housing, transport, urban policy, environment. They may have impinged on the statutory land-use planning system, but only in an indirect way.

But the obverse is not true: planning *did* impinge, deeply and all-pervasively, on those other topics. One senior official used to claim that the DOE Planning Directorate had more influence on government housing policy than the housing people did; there was only a slight degree of hyperbole there, for planning guidelines shaped the entire pattern of national housing provision. Exactly the same could be said of urban policy, where planning policies – or the lack of them, in the enterprise zones (EZs) – had a significant impact on the rate and nature of regeneration.

1970: THE CREATION OF THE DOE

That point is relevant to the origins of the Department. An excellent (but unpublished) paper from Paul McQuail, Deputy Secretary for planning until his retirement in 1994, reminds us that they lay in a very curious arrangement in the last year of the Wilson government, 1969–70, when Tony Crosland became Joint Secretary of State for both Local Government and Regional Planning. As announced by Harold Wilson on 5 October 1969, this was to 'co-ordinate the work of the Ministries of Housing and Local Government and Transport', with special responsibility for implementing the reform of local government in England following the Redcliffe–Maud report, and for the regional economic planning councils created in 1964–5. That last provides one clue: in the same statement, Wilson announced the abolition of the Department of Economic Affairs. It had fallen foul of Treasury jealousy, of course; but the old Ministry of Housing and Local Government (MHLG) was concerned that the councils' work had become physical rather than economic in character, a view very much reinforced by a major struggle in 1967 over responsibility for a new strategic plan for the South East.

Apart from that, there was a strong feeling at the time that strategic land use and transport planning needed to be integrated: a view fashionable in academia, and fortified by the early management of the Centre for Environmental Studies, by the integration of planning and transport during the preparation of the Greater London Development Plan (GLDP), and by the report by Evelyn Sharp – recently retired as MHLG Permanent Secretary – on *Transport Planning: The Men [sic] for the Job*, which changed character in the course of implementation and resulted in the School of Advanced Studies (Ministry of Transport 1970). A critical factor, doubtless, was that Baroness Sharp had been persuaded of the need for a unified department; and she was fortified by an interdepartmental report of May 1969, written by Sir William Armstrong. As Paul McQuail's memoir shows, this latter seems to have reached its conclusion rather reluctantly, stressing the 'very substantial objections' and the 'serious doubts about its viability and manageability'. After a further review and a general election, it was Mr Heath who accepted the recommendation in favour of the new maxi-department.

1970–1975: THE GOLDEN AGE OF SYSTEMS PLANNING AND ITS AFTERMATH

The context is very important, for 1970 was perhaps the high-water mark of a certain style of planning, sometimes called Systems Planning. This was rather heavily influenced by then-current American ideas in management and indeed military strategy, heavily rational in its approach, and fascinated by the new possibilities for modelling and analysis offered by the early computers. It was the spirit embodied in the sub-regional studies which the old MHLG had encouraged in the late 1960s: Leicester–Leicestershire, Nottingham–Derby, Coventry Warwickshire–Solihull, and South Hampshire. And, above all, it was represented by the massive six-volume strategic plan for the South East published by the MHLG in 1970 (South East Joint Planning Team 1970). Further, it represented also the spirit of the new structure plans introduced by the 1968 Act and the expected reorganization of local government on a city–region basis. The expectation was of a very highly articulated system of regional strategic plans and city–region structure plans, which in turn would be embedded within corporate management plans for the entire authority – another major concern of the new Department, reflected by the McKinsey local-authority management reports of 1972 (DOE 1972). Since all these were to integrate land-use and transport planning, the logic seemed inexorable that they should be combined at national level also.

All too soon, this logic began to look rather ragged. First, the South East plan began to founder on local objections; NIMBYism as a word had not been coined, but the reality was already there in the shire counties of the South East, where strong resistance soon emerged to any development that impinged on the countryside. The acid test came in the plan's Area 8, Reading–Wokingham–Aldershot–Basingstoke, one of the major growth areas in the plan, where an attempt was made to forge a sub-regional plan through a consortium of local authorities. Their report of 1975 (Berkshire C. C. 1975) achieved the rare feat of being both indecisive and highly controversial. The resulting arguments rumbled through the late 1970s, with a review of the strategic plan commissioned by the Labour government and published in 1976, and a response from that government in 1978 (South East Joint Planning Team 1976; DOE 1978); it went on into the 1980s, where Michael Heseltine's approval of a relatively modest extension to Wokingham was immediately pilloried under the name Heseltown. Second, Peter Walker as first Secretary of State of the new department took the major decision to reject the city–region solution, enshrined in the 1969 Redcliffe–Maud report (Royal Commission 1969); so the case for an integrated solution at that level was likewise weakened. Third, this period also saw a profound reaction against large-scale integrated transport and land-use planning, both at metropolitan scale, where Walker's successors – Geoffrey Rippon, then Tony

Crosland – had the job of handling the Layfield panel's report on the Greater London Development Plan (DOE 1973), and also at local scale, in Covent Garden, where Rippon in effect rejected the plan by his listing of individual buildings (Anson 1981).

One might also include Rippon's effective rejection of the Travers Morgan study of the London Docklands (Travers Morgan 1973), bequeathed to him by his predecessor, and his decision to go instead for a community-based approach through the Docklands Joint Committee. And finally, Crosland's dismissal of the proposals for the third London airport at Maplin, off Foulness Island on the North Sea coast of Essex, represented a massive rejection of the rational approach to major decision-making, and above all of the use of cost–benefit analysis in planning. The strands came together, in particular, in a pretty severe intellectual onslaught on the leadership of transport engineers in the planning process, as represented by the seemingly inexorable computer models that appeared to prove the need for fourteen-lane motorways, even in the early 1970s; and by the first conscious attempt to scale down the roads programme and to abandon substantial motorway plans for the conurbations, well represented in Tony Crosland's tenure as Secretary of State from 1974 to 1976.

1975–1979: THE DEPARTMENTAL SPLIT AND THE BIRTH OF INNER-CITY POLICY

These decisions, between 1973 and 1976, represented a major sea-change in planning style at every level of government: in essence it was a rejection of the comprehensive, top–down, expert-led, technique-dominated style of the 1960s and its replacement by something like its opposite: heavily community-based, represented by barefoot planners who would serve the needs of the real people against large-scale physical change. It coincided with the publication of the Club of Rome report in 1972 (Meadows *et al.* 1972); the arrival of a Labour administration in County Hall in 1973, and the consequent abandonment of the GLDP roads proposals while still under consideration by Layfield; and by the arrival of a Labour government in 1974, following the great energy crisis of winter/spring 1974 and the associated first great 'Winter of Discontent'. One might think that it had something to do with party politics. But, though Labour in opposition at County Hall took a certain role, the change was interestingly in large measure independent of political ideology, as one can see in a certain continuity of philosophy between Rippon and Crosland; it seems to have represented a general shift in *Zeitgeist*, observable in other countries.

The most dramatic result was the separation of Transport into a separate department, with its own Cabinet minister, in 1976. As Paul McQuail's memorandum shows, there was a political logic: James Callaghan had just replaced Harold Wilson as Prime Minister, he needed to reshuffle his Cabinet following Roy Jenkins's resignation, and he needed to find a Cabinet post for William Rodgers who would become Transport minister. But behind this, clearly, was a feeling on the part of Transport officials that the marriage had not been a happy one for them: perhaps as a result of Labour's election in 1974, far more because of the general shift in perception, there was a major shift in expenditure towards housing and away from transport. In 1970–71 transport took 31 per cent of DOE funding, housing 41 per cent; in 1979–80 the projected figures were transport 26 per cent, housing 49 per cent. Crosland's widow, Susan Barnes, in her memoir about him, recalls his view that it was 'a crude and vulgar concession to the transport lobbies' and that 'It was Jim in one of his irresponsible moments' (Crosland 1982). Although in general senior officials seem to have thought that the combined department worked well, politicians were divided: Walker, Crosland and Heath shared this view, while Rippon and Rodgers were against the principle of a mega-department – perhaps, again, representing the mood of the time that small was beautiful.

Apart from that, the major planning-related initiative of this period was the Community Land Act of 1975 and the associated Development Land Tax of 1976, representing the third attempt by a Labour government – the first accompanying the original 1947 Act, the second in 1967 – to capture development gains for the community. On both occasions, the primary objective was to restore the undoubted logic of the 1947 Act: that, since the nationalization of development rights under that Act and the introduction of comprehensive development controls, gains in land values were essentially generated by the planning system and the resulting gains were public property (cf. Reade, Chapter 5 of this volume). This time, the solution was carefully crafted so as to ensure that the scheme was administered by the local authorities, who would share the resulting gains both with each other – a kind of equalization scheme – and with the Treasury, so ensuring that the latter – Tory as well as Labour – would resist any attempt at repeal. But the passage of the Act coincided with a major slump in the property industry from late 1973 onwards, following the collapse of the late 1960s boom; public expenditure cuts, following the IMF crisis in late 1976, made it impossible to find more than minimal sums of money to provide the essential pump-priming. So, despite some enthusiasm from the property-development industry for retaining the scheme under a Tory government, Michael Heseltine promptly removed it from the statute book in his first legislation of 1980.

The other major influence was indirect: it was the Inner Cities White Paper of 1977, following publication of the three major consultants' reports on Lambeth, Birmingham and Liverpool commissioned by Peter Walker as early as 1972 (DOE 1977a–e), and the bundle of policies that resulted, not least through the Inner Urban Areas Act 1978. This represented one of the biggest single shifts in the whole bundle of regional and urban policies to have occurred in the half-century since 1945. Its effect was twofold. First, even before passage of the 1978 Act, it transferred the existing urban programme from the Home Office, where it had resided since its inception in 1968, to the DOE, thus recognizing that it was no longer perceived as a matter of community or race relations but rather as a multi-faceted programme of urban regeneration, at bottom economic. Second, because the new programme was bigger than the old, the money was found mainly by a sharp shrinkage of the new towns and expanding towns programmes. The latter was effectively phased out quickly, while the former was allowed to proceed to term, in effect bringing the Mark Two New Town Development Corporations – created in the late 1960s – to an end after approximately a twenty-year life during the mid- and late 1980s.

But it can be said that there was a kind of longer-term depth charge contained in this policy shift: effectively it meant the start of a move away from broad-based regional policy – retained, then and later, in the Department of Trade and Industry (DTI) – and into much more closely targeted assistance to smaller urban areas, defined in the original rubric as Partnership and Programme areas. And, especially after the cutting back of DTI regional assistance in the early 1980s, this new emphasis expressed itself even in the regional aid map itself – as in the latest round of changes at the end of 1993, when relatively small parts of London – in Thames Gateway and in Park Royal – were for the first time given Assisted Area status.

The relationship of all this to planning was of course indirect. Shifting resources from new towns to inner-city regeneration meant not merely a geographical shift but also a switch in content, since the new towns had been examples of comprehensive and positive physical planning while the urban programme, at least in its early years, covered a wide spread of economic, social, recreational and physical planning activities. Further, in the original prescription under Peter Shore – Secretary of State since the 1976 breakup – the policies were to be developed by area-based partnerships involving central and local government, thus entirely lacking the strong top–down approach that a development corporation was bound to bring. So there was an irony: despite initiatives like the London Docklands Joint Committee's (DJC's) 1976 Docklands Strategic Plan, the change was perceived as a movement away from a strong planning stance. Indeed, between 1976 and 1979 the DJC came to be widely criticized for lack of effective action to regenerate its area, though its defenders later claimed that it performed much groundwork for projects, such as housing in the Surrey Docks and at Beckton, that later accrued to the credit of the London Docklands Development Corporation (LDDC).

In general, the Callaghan government can be seen as a curiously transitional one in late twentieth-century British history: the IMF crisis, and the savage public expenditure cuts that resulted, played a role rather like the very similar crisis of 1967 which had caused the Wilson government to jettison the National Plan. But the effects were compounded by the deepening recession and the emerging evidence of structural economic crisis, particularly but not exclusively concentrated in the inner cities. As a result, virtually all major reforms were put on hold and it proved almost impossible to implement the manifesto commitments, notably that on land. In some ways, indeed, the period can be seen as a precursor for the radical changes that came after 1979. Particularly, Callaghan's very public abandonment of Keynesian macro-economic policies provided a logical basis for the much more severe public expenditure cuts that Margaret Thatcher carried through the deepening recession of 1979–82.

The era provided other policy precursors. One, interestingly tucked away at the end of December 1977, announced that proposals for hypermarket and superstore proposals would be considered in terms of the existing patterns of shopping while taking into account adequacy, convenience and the need to retain the vitality of town centres. It used a phrase that would be used again and again, with slight variations in emphasis, by Conservative ministers in the decade that followed: that it was not the purpose of planning policy either to prevent or to stimulate competition among retailers and types of shopping. This statement marks the beginnings of a marked policy shift, the full importance of which came to be felt only in the late 1980s and the 1990s, as a flood of new superstores on non-traditional, non-central sites came to threaten the viability of town centres.

1979–1983: THE PERIOD OF TORY CORPORATISM

May 1979 saw what, by general agreement, was the most radical break in British politics since 1945. But it is significant that, in important respects, the break was in the means to achieve defined policy ends, not in the ends themselves. Thus DOE officials could say, in their 1985 evidence to the Church of England Commission, that 'the present Government largely accepted the analysis in the 1977 White Paper'; the difference lay not in the definition of the problem but in identifying the solution. Labour saw the means as co-operation between central and local government, with private enterprise as a shadowy element to be either confronted, or at best negotiated with; the Conservatives under Heseltine replaced this with a partnership between central government and business, with local government marginalized.

Behind that lay a difference of view not only as to the most effective mechanism, but also as to policy content: Heseltine saw Labour city councils as wedded to outmoded and dying economic activities, and was determined to find a way of recycling underused land to accommodate the new informational jobs of the post-industrial economy and also the owner-occupied housing of those who would work in them. Hence, not only his preference for the Urban Development Corporation – ironically, quite openly modelled on the Attlee government's formula for the new towns – but also his repeated obsession, almost from his first day, with local-authority land registers and the release of urban land for the new purposes. This very neatly combined a policy initiative with a political advantage, for the policy would put new Tory-voting electors into old Labour bailiwicks as well as taking unpopular new development out of NIMBY shire counties.

Again, as under Callaghan, the most important of these initiatives lay in the field of urban policy, not of planning. Started soon after the 1979 election, such initiatives were of course given a boost by the riots of spring 1981 and by Heseltine's seizure of the policy initiative through his month-long visit to Liverpool that summer. But there was a relationship between urban regeneration and planning, provided by the new emphasis – already underlined in Geoffrey Howe's first budget – on enterprise zones (EZs). The first EZ was launched at Corby in July 1981; by November ten out of the first eleven zones were operating. They had originally been seen in terms of almost planning-free areas as a desperate final attempt to regenerate the most difficult cases of de-industrialization; but in practice their attraction to investors largely lay – as subsequent research showed – in straight fiscal incentives, principally a ten-year holiday from local property taxes and from corporation tax. Relatively rarely did EZs form part of UDC areas; the Isle of Dogs was a rather special exception, which showed that a combination of the two policies in the right location could achieve rather remarkable outcomes, almost certainly beyond anyone's original expectations. However, in important respects the package was no different from locating a 1940s new town, such as Peterlee or East Kilbride, in a development area; it demonstrates the remarkable continuity of policies rather than the reverse.

It is true that planning featured a great deal in the early policy statements of Michael Heseltine, almost as a villain of the piece. Planning, he already said in September 1979, was a contributory factor to British decline; he had no intention of scrapping the system, but he believed it tried to do too much, and above all he wanted to speed up its operation at every stage. Thus, structure-plan preparation should be greatly speeded up, in particular by dropping unnecessary survey work; local plans should be produced in advance of structure plans; and development control should be hugely accelerated, with some applications entirely freed from development control, most cases decided within eight weeks, and with no attempt to second-guess architects on matters of taste: Heseltine referred contemptuously to squaring the dome of St Paul's, a phrase that John Gummer might well have used in 1995 when he defended variety in architecture, even when it offended canons of conventional good taste. And finally, appeals should be greatly streamlined and speeded up, with all cases resolved by inspectors.

In fact, much of this was done. By November 1982, land registers were published for the whole of England; the counties were pressured into bringing their structure plans forward, and by February 1982 only one out of eighty-two was outstanding; changes in the General Development Order freed small extensions, especially residential ones, from development control altogether; the percentage of applications determined within eight weeks rose from 61 to 69 between the second quarter of 1980 and the second quarter of 1981, subsequently rising to just over 70 per cent in 1982-3 before sinking catastrophically in the late 1980s boom to as low as 52 per cent in April–June 1990; all categories of appeal were transferred to inspectors as from July 1981, and the procedures were speeded up.

These essentially were the fruits of Heseltine's tenure as Secretary of State, and of Thatcher's first term. Political commentators and academic political scientists have wrestled with the problem of the political philosophy that Michael Heseltine represented in this period, and have come to the interesting conclusion that it was a kind of radical corporatism. Margaret Thatcher was right, they conclude, when she said that he was not 'one of us': he shared with her the belief that government must be slimmed down so as not to obstruct the enterprise culture, but at the same time he actually wanted to extend government activity into the crucial job of urban regeneration (Critchley 1994). There was a close parallelism with Peter Walker's approach on London Docklands in the first years of the DOE, and it was perhaps no accident, even poetic justice, that Walker should have been instrumental in setting up the Cardiff Bay UDC, subsequently becoming its chair. It represents a strand of Toryism that was surprisingly persistent throughout the Thatcher era, and made a very definite reappearance in Heseltine's second manifestation at the DOE in the early 1990s.

1983–1987: THE RADICAL ATTACK ON PLANNING

After a short period in early 1983 during which Heseltine was replaced by his planning minister, Tom King, following the general election, Patrick Jenkin came in as Secretary of State, and with him came a sharp intensification of Thatcherite Tory radicalism. It was during the years 1983–7, first under Jenkin and then – following a short period under Kenneth Baker – during Nicholas Ridley's occupancy of the office, that the philosophical attack on the planning system reached its apogee. First, following a manifesto commitment, the White Paper, *Streamlining the Cities*, asserted that there was 'no role for so-called strategic authorities; created in the 1960s and 1970s, it is now clear that they result in friction and duplication'; it therefore proposed to abolish the Greater London Council (GLC) and the six metropolitan counties in the major provincial conurbations (DOE 1983). After a ferocious political battle lasting more than two years, abolition indeed occurred on 1 April 1986. Not the least important aspect was that with them went any overall strategic or structure plan for these seven areas, where subsequently there would be a single-tier plan system consisting of Unitary Development Plans (UDPs) for each London or metropolitan borough council. Critics, not just Labour ones, pointed to a resulting anomaly: that structure plans were missing in just the areas where the problems of interrelationship and co-ordination were greatest, while they survived in the rest of the country. And, though the resulting gap might be filled by regional strategic guidance from the DOE, the statements in the early and mid-1980s were habitually so laconic – as in the famous two-page South East guidance in Michael Heseltine's time – that they did not fulfil the role.

In July 1985, when Baker had succeeded Jenkin as Secretary of State and the GLC–metro abolition was going through its final throes, the White Paper *Lifting the Burden* took the logical step of proposing simplification and eliminating duplication in the planning system; a year later, June 1986, a month after arrival, his successor Nicholas Ridley published a consultation paper, *The Future of Development Plans*, proposing a single-tier system of planning in the rest of the country, structure plans effectively being abolished and replaced by a much broader and vaguer system of county review (Minister without Portfolio 1985; DOE and Welsh Office 1986).

In parallel with this reform of machinery, ministerial announcements during the mid-1980s demonstrated a progressively more radical anti-planning tone. Already, in a speech to the Royal Institute of Chartered Surveyors in May 1984, Patrick Jenkin had unveiled his plan for simplified planning zones, modelled on the enterprise zones' planning procedures; local planning authorities could if they wished take the initiative in granting advance permission for any development in parts of their areas, without the need for planning applications. They required legislation, and were eventually embodied in the Housing and Planning Act 1986, finally coming into force only in November 1987; they have had curiously little effect in practice. A year later, in a speech to county planning officers in March 1985, Jenkin was calling for 'more market-oriented planners' and stressing that the presumption must be in favour of development save where there would be demonstrable harm – for instance, development on green belts, in areas of good agricultural land, or in spaces between neighbouring towns. He referred to the intense pressures that were building up in the South East and the strong NIMBY pressures there – though this term was first used officially by his successor Nicholas Ridley. In July of that year, he was emphasizing that it was not the government's function to limit the competition that large new stores would offer, save where they threatened the vitality and viability of centres as a whole – a phrase that was to be reiterated many times. By that point there were already 160 hypermarkets and superstores as against 125 in January 1979 and a mere 26 in January 1973; and the impacts on town centres were just beginning to emerge.

The radical Tory campaign seems to have reached some kind of high-water mark at the end of 1986, possibly because a general election was beginning to loom. Already in January 1987, in response to press reports, Ridley was affirming that there was 'no secret plan to abolish planning control'; his Planning Minister, William Waldegrave, was affirming that one-tier planning represented no threat to the countryside, and that counties could still take the lead. By April 1987 Waldegrave was actually asserting that regional strategic guidance was vital, and was commending SERPLAN, the Standing Conference of South East Planning Authorities, as a model for the rest of the country. This was a key announcement, for it presaged what was to become the most significant development of the late 1980s, the systematic cycle of regional advice and regional guidance which SERPLAN had helped to pioneer.

But most significant of all was the fate of the proposals for privately developed new communities. A consortium of the country's largest volume builders had organized themselves in 1983 into Consortium Developments with an ambitious programme to develop new country towns in the South East. Their first attempt, a scheme for 5,100 houses at Tillingham Hall near Thurrock in Essex, was perhaps ill-judged, possibly a deliberate attempt to test the water: it was in the middle of the metropolitan green belt. In February 1987 Nicholas Ridley accepted the inspector's recommendations and rejected it. It caused little surprise within the planning profession; but it was to be the first of a whole series of such reverses for the volume builders.

1987–1995: THE REMAKING OF THE PLANNING SYSTEM

These reversals might be dismissed as the effect of pre-election nerves. But in fact they presaged a major shift in policy. Some of the developments, needless to say, continued along lines that by now had become traditional. Thus in February 1988 there was an announcement about the new Unitary Development Plans that would replace the old two-tier system in London and the old metropolitan county areas; in June yet another attempt was made to speed up planning inquiries; most significant of all, in October of that year Michael Howard, Planning Minister in succession to William Waldegrave, introduced the new General Development Order which included general permission for change of use between factories and other business premises, to become effective in December. It had a significant effect in allowing changes of use on large areas that clearly would not be used for industry again; but perhaps even more significant was that local authorities began to show much greater flexibility in giving permission for more radical recycling: superstores on old railway sidings, multiplex cinemas on old factory or showroom sites.

Nonetheless, from the start of 1988, with Nicholas Ridley still Secretary of State, there emerged a series of initiatives which progressively reshaped the planning system. The first, right at the start of the year, was the announcement of a new series of Planning Guidance Notes and Mineral Guidance Notes, intended to be more comprehensive and also clearer than the old Circulars which had emerged in such a steady stream ever since the 1947 Act.

It was accompanied by the issue of Planning Policy Guidance Notes (PPGs) 1-9 and Mineral Policy Guidance Notes (MPGs) 1–2. Ridley did make one last desperate dash for freedom: in January 1989, following a general government White Paper, *Releasing Enterprise*, in November 1988 (Secretary of State 1988), he proposed another attempt to simplify the system with a single tier of plans in the shire counties, the county role being reduced to policy statements; but in October, after a consultation exercise that brought very fierce reactions, his successor Chris Patten decided against any attempt to relax planning controls in the countryside. Nonetheless, the Planning and Compensation Bill – introduced in November 1990, and enacted the following July – did provide for a streamlined structure-plan process.

This was accompanied by a further retreat on the issue of new countryside settlements. In the summer of 1989 Ridley rejected the proposal for Stone Bassett east of Oxford on the ground that it was contrary to the Oxfordshire Structure Plan, but he announced that he was 'minded to allow' a proposal at Foxley Wood between Reading and Basingstoke. Almost immediately on succeeding him, Chris Patten in October 1989 announced that he was minded to reverse this decision, and in December he confirmed this, stating 'We are not in the business of sacrificing environmental quality to sheer housing numbers', thus effectively backtracking on Ridley's implication that there was such a strong case on housing grounds that other considerations should be set aside. In February 1990 he turned down another nearby proposal, Great Lea, south of Reading; the entire policy was now in reverse, assuming that it had been pointing forward in the first case. Patten may have felt that the pressures for new development, so strong during the late 1980s, were at last weakening: later that year it was reported that applications for the April–June quarter had dropped by 17 per cent compared with the previous year, demonstrating that the recession in the development industry had well and truly begun. By 1992, the circle had evidently turned: the new PPG 3 on *Housing*, issued in March, stated that new settlements should be contemplated only where the alternative of extending existing towns and villages was less satisfactory and where there was clear local support (DOE 1992a).

The most important legislative developments in this era were the Town and Country Planning Act 1990 and the Planning and Compensation Act 1991, which effectively form a whole. Not only did the 1991 Act provide for counties to adopt their own structure plans without the need for approval by the Secretary of State, a considerable simplification; it required district authorities to prepare local plans for the whole of their areas, required that they must take account of environmental considerations and gave them greater powers to insist on environmental assessment of development proposals, standardized plan-preparation procedures, and improved and streamlined enforcement powers. But, most significant of all, following Section 54A of the 1990 Town and Country Planning Act, it required planning decisions to accord with the development plans unless other material considerations demanded otherwise.

In autumn 1991 and spring 1992, soon after the passage of the Act, planning minister Sir George Young began to emphasize the significance of the provisions that took effect the following February: we had begun to move to a plan-led system, in which a hierarchy of government advice – first the PPGs and MPGs, then the RPGs or Regional Planning Guidance prepared on advice from regional consortia of planning authorities – would provide a clear framework for Unitary Development Plans or structure plans accompanied by area-wide district plans; these in turn providing the definitive evidence for resolution of development proposals or appeals. It was not, he stressed, a 'bombshell' against development; indeed, the presumption remained in favour of development, as it had ever since the General Development Order of 1948; only rarely could there be a presumption against development. This was followed by a direction to all planning authorities that they inform the Secretary of State of all applications involving a departure from the development plan, so that he could decide whether to call them in. In September 1992, the government announced that it wanted to achieve substantial district-wide plan coverage by the end of 1996; it was a daunting target, but the Section 54A provision gave local planning authorities a built-in incentive to meet it.

So much for the legislative framework. In terms of policy, there were at least four major initiatives. First, at the strategic planning level, came Michael Heseltine's East Thames Corridor announcement of March 1991, only four months after his return to the post of Secretary of State: in effect a massive eastwards continuation of the London Docklands Project over some thirty miles of the lower Thames valley and estuary, but now emphasizing a combination of greenfield and brownfield development rather than pure urban regeneration; and, associated with that, the subsequent establishment in 1994 of English Partnerships, the English Development Agency which Michael Heseltine had proposed from political exile in 1987. Thames Gateway, as it was subsequently renamed, was to be developed not through an Urban Development Corporation but through a broad sub-regional framework developed by an internal task force, set up in March 1993, and then through area-based strategies developed by consortia of local authorities and developers; the agreed framework was published in spring 1995 and the first major developments, at Barking Reach and in Thames-side Kent, were already beginning to take shape.

Second was the new emphasis on sustainable urban development, based on the research by ECOTEC and enshrined in PPG 13 on planning and transport published in draft in April 1993 and finally in 1994 (Department of the Environment and Department of Transport 1994). This document was in many ways a significant policy initiative, not least because it resulted from long negotiations with the Department of Transport, themselves engaged in a policy shift away from a strong roadbuilding emphasis toward a more balanced package with greater emphasis on public transport, walking and cycling. Its publication was followed in July 1993 by a significant event: the two departments announced that they had abandoned their plan for the East London River Crossing (ELRC) because of the impact of the link road through Oxleas Wood, an area of ancient woodland in south-east London. This followed intensive protests by anti-road campaigners against the M3 extension at Twyford Down near Winchester and the Hackney–M11 link road through north-east London, and appeared to signify a major policy shift; particularly since ELRC was a major element in the Thames Gateway strategy. Clearly for this reason, the statement announced that alternatives would now be studied; as at the time of writing (October 1996), no firm announcement has been forthcoming. PPG 13 illustrated how the two departments, divorced for seventeen years and no longer even enjoying common citizenship, could at last come together to forge a policy document on a subject that involved exceptionally complex relationships between land-use and transport policies: the very argument for the marriage of 1970. It was followed in July – a week after the ELRC announcement – by another significant initiative from DOE, but with important transport implications: the publication of the draft of the revised version of PPG 6, the first major statement on town centres and retail development to come from the department in twenty-five years (DOE and Department of Transport 1993). It stated that the aim of policies on new stores should be to provide a wide range of opportunities for all. It reiterated a statement made many times in previous statements, that applications for superstores should be rejected only if they threatened the viability and vitality of town centres as a whole. However, though in preparation long before, it happened to be published two months after the arrival of John Gummer as Secretary of State; soon after, in October and November 1993, Gummer began to make a series of major statements that indicated his clear personal determination to end the era of *carte-blanche* for out-of-town and edge-of-town developments. It was followed in July 1995 by a new and even tougher draft of PPG 6, with a more stringent set of tests for permissions on new stores, including access by a choice of different forms of transport, and impact not only on vitality and viability, but also on overall travel and car use; this and PPG 13 both followed the clear lead that had been set out under Chris Patten, in 1990, with the department's environmental statement, *This Common Inheritance* (DOE 1990). Overall, there was a clear indication that city-centre sites would be preferred from now on; and there was also a new emphasis by John Gummer on the virtues of mixed-use schemes that combined housing, shopping, services and work opportunities (DOE and Welsh Office 1995). Cynics were not slow to point out that, with some 400 schemes with planning permission in the pipeline, this was a case of shutting the stable door after the horse had bolted; a view reinforced in early 1995 when it was announced that

Australian finance had been found to start work on one of the country's largest edge-of-town regional centres, Bluewater Park in Dartford.

Bluewater represented a major dilemma in the department's strategic policy in the mid-1990s. On the one hand, it reflected the ideas of the mid-1980s and was completely contrary to current guidelines; it also represented a substantial invasion of the metropolitan green belt, on which the department issued further and stringent policy guidance in October 1993 (DOE and Welsh Office 1993). On the other hand, it was a key element in the Department's own framework for strategic planning in Thames Gateway, published in June 1995, and in the sub-regional framework for Thames-side Kent that followed in September (Thames Gateway Task Force 1995; Kent Thames-Side 1995).

Third was the announcement in November 1993 of integrated regional offices for each of the English regions plus London and Merseyside, bringing together for the first time the work not only of the Departments of the Environment and of Transport, but also of the Department of Trade and Industry and of Employment (before its merger with Education). In some ways this was potentially the most important development of all, almost harking back to the spirit of integrated regional planning that attended the creation of the original Ministry of Town and Country Planning in 1943. It represented a major shift in the pattern of government in the regions, which would take time to bring into effect and even longer to evaluate properly; the offices duly came into formal existence on 1 April 1995, though physical integration has in some cases taken longer. And critics were not slow to point out that this represented a further invasion of central power downward rather than the extension upward of local power into the regions, which the Labour Party was tentatively embracing.

Finally, John Gummer gave his personal imprint to a new initiative to improve the quality of town and country: in some ways an echo of Michael Heseltine's emphasis in the early 1980s, but now given extra point by frequent media comparisons with the Parisian *Grands Projets*, and also the new opportunities offered by the millennium funds. It bore fruit in a consultation paper, *Quality in Town and Country*, issued in December 1994 (DOE 1994c); in a consultants' report on strategic planning guidance for the River Thames, published by the Government Office for London in April 1995 and deliberately aimed at improving the quality of the Thames-side landscape in London (Government Office for London 1995); and by new plans from the South Bank Centre for the refurbishment of the South Bank. Interestingly, there was a heavy architectural emphasis in these schemes, underlined by Richard Rogers' Reith Lectures in the winter of 1995: there is a new interest in urban design at the local level, influenced perhaps by Prince Charles's strictures on the failures of planning, but also – cynics again might say – by the lack of architectural commissions in the early 1990s recession.

POSTSCRIPT, 1996: FORWARD TO THE MILLENNIUM

One could fairly conclude that planning survived the ideological onslaught upon it in the mid-1980s; it is strongly established in the agendas of both New Toryism and New Labour, because at the end of the day it has very firm support in the political agenda of the voters of Middle England. The system forged from 1987 onwards – planning guidance, regional guidance through integrated government offices, structure plans and UDPs, area-wide district plans, plan-led development control – is a system as coherent as any that has existed since the historic 1947 Act. Fortunately, also, it is flexible enough to absorb shifts in policy arising both from political ideology and changing circumstances.

There are, however, some key remaining questions for the last five years of the century. The first and most obvious concerns local-government reorganization. We are emerging with a variable geometry: a variant of Peter Walker's two-tier system in most of England, but with a return to all-purpose old-fashioned county boroughs in many, and with a few counties carved into all-purpose unitary authorities. There are going to be real problems of organizing structure planning in these fragmented areas, and it will need a great deal of goodwill across boundary lines, and in some cases across political boundaries, to achieve it. Related to this is the promise from the Labour Party, if indeed it does remain a promise, to move in stages from devolution in Scotland and in Wales toward a regional system of government in England. Such a system could comfortably absorb the government offices and would conform with the new structure of a Europe of the Regions, which is one of the less well-recognized outcomes of Maastricht. But it raises many questions: which functions would be abstracted from central government and which from the local authorities? Could the system of regional advice and regional guidance then be collapsed into one? And, not least, what would be the precise geography, particularly in and around the conurbations?

The second question is perhaps the most acute for the next decade: it is how to reconcile any notion of sustainable urban development with the 1992-based household projections, published in autumn 1995. For they suggest that in the next twenty-five years, 1991 to 2016, we shall have to accommodate 4.4 million additional households: 3.7 million in the twenty-year period down to 2011. No less than four in five are expected to be one-person households, the product of more young people leaving the parental home for higher education or first job, more divorces and separations, and more widows and widowers surviving their partners. Further, 1.7 million of the 4.4 million, over one-third, are expected to be in the South East, already the most pressured region in the country (DOE 1995).

The DOE's policy position is that across the country, no less than 60 per cent of all housing should go within the urban boundaries, on so-called brownfield land: land left waste, old industrial or railway land, or residential areas redeveloped at higher densities. Rather remarkably, by the early 1990s almost exactly that target was being reached. Some bodies would like to see that figure go even higher. But an inquiry by the Town and Country Planning Association (TCPA), for the Joseph Rowntree Foundation, published in summer 1996, concluded that even the previous 50 per cent target would most likely not be maintained: the brownfield land was likely to dry up, and any still waiting for development might be prohibitively expensive (Breheny and Hall 1996). The challenge, it suggested, would be to devise portfolio approaches for each region, combining maximum possible urban compaction with a variety of solutions to sustainable greenfield development – above all, through mixed-use developments along public-transport spines.

Clearly, regional guidance from the new integrated government offices would play a key role in this. But the same report found that, in the regions, representatives from the counties and the districts complained that the guidance was often bland and unadventurous, failing to give a strong lead to the good-quality development that was needed. For the future, it seems, there will be a need for regional standing conferences of local planning authorities to work closely together with the regional offices to produce more positive guidance. And, under a Labour government prescription, the regional offices themselves would become part of a regional government structure. Whether that would make it easier, or the reverse, may be discovered in practice

The related question is whether all this can be done in a very short time without compromising urban quality – a key theme of John Gummer's entire period of tenure at the DOE. The experience during the previous period of rapid urban growth in Britain, the 1960s, does not augur well: this was the decade during which desperate local authorities, concerned with speed at the expense of everything else, commissioned off-the-shelf industrialized building packages that too often proved disastrous. True, that was in an era of large-scale council housebuilding that has long since gone; but, as the TCPA report suggests, if the least fortunate members of society are not to lose badly in the coming scramble for housing, it may now be necessary to expand a ring-fenced social-housing programme. The need then is for an orderly programme of land banking and land release to take account of long-term needs, while adapting the precise rate of release to year-by-year changes in pressure, and while making specific provision for social housing, however that term is defined.

This is a formidable agenda. And it brings planning and housing back into the kind of close relationship that obtained in the 1950s and 1960s, for very similar (though not identical) reasons; so it is centrally a DOE agenda. Further, since it links to sustainable development and thus to PPG 13, it is also a joint Environment/Transport issue. Maybe this illustrates the unwisdom of splitting the department in 1977, and the possible wisdom of rejoining the two halves twenty years later. Certainly, with so much everyday implementation in both departments hived off to agencies such as English

Partnerships or the Highways Agency, the case for remarriage is stronger than it ever was in most of the intervening years. But it also points to an intrinsic organizational problem within DOE: the links between planning and urban regeneration on the one hand, planning and environment on the other, are so strong and so complex that crucial issues like the development of Thames Gateway, or the search for sustainable forms of urban development, will necessarily involve cross-relationships between DOE policy directorates.

There is a tendency here to shrink planning to the narrowest definition, and to treat all the related areas as quite separate. The plan-related system, I fear, has contributed to this narrowing of the horizon: planning comes to appear as ever more routinized, and therefore boring; the interesting and creative jobs are stripped away and put elsewhere. Significantly, in the reorganization of late 1995, Thames Gateway was taken out of planning where it had previously belonged, and put into the portmanteau urban and rural directorate. And parallel to this, there is an unfortunate tendency for the professional planning element in the Department to shrink in both size and influence. Consider: at the formation of the DOE in 1970, the professional planners were headed by a Grade 2, Sir Wilfred Burns; there were three deputy chief planners at Grade 3. In the mid-1990s the head of profession is a Grade 5; the post of chief planning adviser is a half-time appointment at Grade 3. And the entire professional staff has spectacularly shrunk in numbers, in parallel with the general decline in the top administrative grades, as staff have taken early retirement and found good career opportunities outside the department. The clear implication is that planning professionalism has no special place or influence and that – as has obtained ever since the 1970s – at the higher levels of policy making there is a single administrative hierarchy. This must reflect the diminished status of planning in the world outside; but it is strange to see it in so stark a form in circles that are presumably better informed.

ACKNOWLEDGEMENT

The author wishes to acknowledge the generous agreement of the Department of the Environment to the publication of this chapter, originally contracted by them for their 25th Anniversary Conference in London, October 1995, and lightly revised for appearance here.

8

TOWN PLANNING INTO THE 21ST CENTURY

Diverse worlds and common themes

Cliff Hague

There are times when an old order is finished but the shape of the new one is not clear. The 1990s is one of these. The passing of the post-World War II era of Cold War and welfare states was protracted. Halsey (1987) suggested that the oil crisis of 1973–4 was widely acknowledged to have ended the post-war period for Britain and for other 'First World' countries. Full employment, economic growth and increasing state intervention were replaced by mass, long-term unemployment, restraints on local-government spending, and a shift of decision-making into the market. The collapse of the Soviet Bloc in 1989 completed the political transformation that the economic changes from the 1970s had initiated. The geo-political order was restructured; internal politics and development processes in the satellite states were transformed, there were no longer two super-powers bidding competitively for influence in the development of non-aligned countries in the 'Third World'.

In the western countries the changes in the political economy have impacted on the role of the state. The state (and the UK is a typical example) has shifted from being a provider of a comprehensive range of public services to an enabling state where the private sector assumes a more pervasive role. The result has been a world of private affluence and public squalor, where deep social cleavages fester. It is ridden with insecurity and uncertainty, even amongst those affluent individuals and countries who are its most evident beneficiaries. The spectres of economic decline, urban crime and environmental degradation stalk the future.

In thinking about 'Town Planning into the 21st Century' at an international level, this chapter looks critically at the legacy of what Evans, in Chapter 1 of this book, has termed 'classical town planning'. Across much of the globe, town planning was one of the building blocks of a now defunct era. In the west it was part of that welfare-state settlement, a means for a public-investment-led reconstruction of the cities undertaken by professionals in the public interest. Behind the 'Iron Curtain', town planning was cast within a system premised on Engels's (1959: 111) notion that, once the proletariat had seized power, 'Socialized production upon a predetermined plan becomes henceforth possible.' Elsewhere the structures and thinking associated with 'classical town planning' were a colonial then a neo-colonial imposition. Everywhere town planning is being challenged and restructured, yet ironically (and irony is the *leitmotif* of the 1990s) there seems to be a convergence towards a restatement of some of the abiding themes and features of town planning as we approach the millennium. Provision of safe and healthy living conditions; the functional efficiency of cities; conservation of built and natural environments; a balancing of public and private interests; effective use of resources – all these concerns that are now to the fore show strong continuities with traditions in town-planning thought. The dualisms on which critiques of 'classical town planning' were built – technical/political, plan/policy, process/ends, professional/ 'community' – are themselves being left behind by the currents of post-modernity.

GLOBALISATION AND TOWN PLANNING

Town planning is typically a local practice set within a distinctive national or provincial legal code. This is why many debates about town planning are essentially parochial, focusing, for example, on local initiatives, or the particular national institutional or policy context. Yet international influences set the parameters of urban change and planning. In the old industrial countries factories close because of shifts in the value of currencies, or because distant head offices rationalise and relocate production to maintain international competitiveness. Confronted with the resulting derelict sites and unemployment, town planners revamp the image of the place, chase European Union structural funds, and 'town planning' blurs into 'local economic development'. The local practice is triggered by, and tailored to, global change.

The opening of the Berlin Wall in 1989 transformed the agenda for town planners practising throughout the vast Soviet bloc. Town and regional planning had been proclaimed practices of the communist regimes, providing the necessary spatial dimension to implement sectoral plans. The embrace of the market and pluralist politics drastically changed the local practice of town planning. New labels, such as 'spatial planning' or 'urban management', have been adopted both to distance practices from the discredited structures of the past, and to recognise that new attitudes and skills are now needed.

Far reaching as these impacts are, it is in the less developed countries that the impact of global forces is most stark and the need to rethink 'classical town planning' is most acute. The town-planning systems in many such countries, as colonial implantations, originated from global influences. Today those systems confront the remorseless movement of people to the cities to live in tents, shacks or on the streets. Behind this impoverished urbanisation stand world commodity markets, the new international division of labour and the mountain of debt to western banks.

All places are now part of the global market economy and are competing against each other with a new intensity. As technological change and the global extension of market forces have ironed out differences in productivity between places, other spatial differences have assumed greater importance, including the quality of life and the image of the place. This puts a new premium on the management of places as entities, over and above the development of particular sites. The traditional forms of town planning may no longer be appropriate, but care, creation and promotion of places in a way that transcends competing private interests is essential to success in the new order of things.

There is a further theoretical and practical conundrum in all of this. We have a body of theory about the nature and working of markets, developed in neo-classical economics, which is remarkable for its robustness, its certitude and its claims to generality. Yet the total globalisation of markets has coincided with an increasing awareness in social science of the very opposites of these qualities. Contingency, fragmentation, discontinuity, uncertainty, non-linearity, difference, locality and a preference for 'weak' theorising stand in stark contrast to the overarching strengths of the abstract and placeless market. This post-modern *zeitgeist* has even infused the popular management literature written for those in corporations and private enterprises whose day-to-day actions build the structures of the global market. Peters (1989) stresses the unpredictability of the business environment, and the extent to which success or failure depends on attitudes. If we follow this logic, then attitudes and values are likely to become increasingly significant influences on the effectiveness of town-planning practice.

KEY INFLUENCES

In so far as the main factors which will influence the progression of town planning into the twenty-first century are fundamentally international in character, they can be identified and discussed in generalised terms. They point towards some convergence in the skills and attitudes required amongst town planners, though multiple contingencies and agencies will fashion the actual impact between different places. As markets become all-pervasive, town planning becomes an exercise in managing change rather than imposing comprehensive designs. Three key influences will steer town planning into the new century, and all three are rooted in the globalisation of the market economy. They are recession, resources and plurality.

When the first draft of this chapter was read at a meeting organised by the

Institute of British Geographers in 1992, it argued that recovery from the international *recession* would be slow. It is now clear that recovery will also be weak. Indeed, recession followed by weak recovery will characterise the era stretching into the twenty-first century in the way that growth and full employment did for the generation from 1950 onwards. The kind of boom that so dominated thinking and politics in the south of England in the 1980s is unlikely to be repeated. The twenty-first century is likely to continue the present negation of the traditional state/market dualism, and be a period in which the central state continues to facilitate the operation of a relatively free market.

Town planning in a period (or a region) in recession has significantly different features than when/where the economy is strong. Not only is there less pressure for development, but an extra premium attaches to prime sites because the risks for investors are fewer there. Plans do not date so quickly in a recession, and are likely to be more appreciated by developers because the greater certainty they provide further reduces risk. However, where there is no development pressure, land-use plans by themselves will not be enough. Recession puts town planners in a weak bargaining position *vis-à-vis* developers who have to be enticed into taking on projects. State or multinational superstate initiatives can play a critical role in easing recession. Therefore there must be a strategic dimension to town planning. Key sites and resources have to be identified, promoted and protected from the uncertainties that a competitive and unplanned market would create. In a recession, town planners need imagination and confidence to see new opportunities, the ability to communicate such possibilities, and skills of persuasion and negotiation to make things happen.

There are two aspects to *resources*, one financial, the other environmental. Because of recession, financial resources will be limited and this will be true for both public and private sectors. Town planners will have to look to mobilise other resources, including voluntary initiatives and intangibles such as commitment to place and the physical quality of an environment. Free markets are corrosive with respect to resources. By restricting value to commodities and then destroying such value to make new opportunities for capital accumulation, markets negate other possibilities. The degradation of so much of the public space in urban areas arises directly from the fact that it is not a commodity. The scope and rewards for voluntary work are largely invisible because the labour involved is not a commodity. Loyalty to place is incomprehensible in a view of the world as a surface differentiated only by rates of return on investment. Faced with these negatives, the traditional agenda of town planning has some contrasting strengths. Efficient provision and use of infrastructure, respect for public space, encouragement of voluntary initiative and the fostering of civic identity are causes that the town planning profession has espoused. They make good sense but will not be delivered by unfettered market forces.

The second aspect of resources is natural resources and their conservation.

The green agenda, which is both international in its ethos and analysis yet also decentralist and participatory, will be with us through into the next century because direct (though not unproblematic) links can be made between sustainability and the quality of life. Of course recession makes it more likely that decision-makers will give priority to jobs and development over environmental conservation, to 'sustainable development' with all its contested meanings over 'sustainability' (Jacobs and Stott, 1992). Nevertheless the need for sustainability will not be diminished by recession, and the creativity and imagination released by the idea have particular value at times when financial resources are tight. Environmental concerns and a conservation ethos are part of the town-planning tradition, and provide a basis for reconstructing town-planning practice up to and beyond the year 2000.

One of the strengths of markets is their capacity to reflect and even to foster *pluralism* and diversity. So many features of post-modern societies that are seen as cultural phenomena were in fact nurtured by market forces. The burgeoning international youth culture is an obvious example of the way a market economy can divine otherness and then generalise it through commodities. Open labour markets have drawn ethnic-minority communities into the cities. The tourist industry has opened up places and their associated cultural experiences to consumers seeking something different to their day-to-day environment. Heterogeneity will therefore be an increasingly significant feature of the cities as we enter the new millennium.

Race and poverty are interlinked dimensions of the increasing plurality of the city – gender, area, disability, age, sexual orientation are others – and illustrate the challenges that town planners will face. Everywhere markets will move the poor into the cities and from poorer countries to richer countries. Minority populations will become more assertive in demanding that town planning and urban policy serve their needs and there will be heightened conflict about urban development. Los Angeles has been seen by some as the archetypal city of the future with its conjunction of the First World and the Third World (see Soja and Scott, 1986). The major riots there in 1992 were caused by poverty and frustration with racist practices by public authority. In 1992 Europe experienced the most massive migrations since the end of World War II as the ideology of 'ethnic cleansing' created a diaspora from the former Yugoslavia. Recession and the end of communism saw a renaissance of fascism in the former East Germany, with ethnic-minority populations on the receiving end of street violence. In Amsterdam, where what is termed the 'foreign population' is expected to increase to 34 per cent by 2010 (Jobse and Musterd, 1992: 53), there were press reports of white residents responding to an horrific airplane crash in the Bijlmermeer (an estate known for its poverty and ethnic minority population) with comments of 'good riddance'.

The risk then is that the weak markets and weak states will be compounded by weak civil societies rent by prejudice and intolerance.

'Classical town planning', as defined by Evans, showed scant sensitivity to pluralism; such planning was in aggregate terms for some standardised perception of needs. Just as the planners' public-interest ethos must be retained but tuned to the realities of societies based on market forces, so that same ethos needs to be revamped to protect the interests of minorities and to deliver equal opportunities.

THE CHALLENGE OF RAPID URBANISATION

More than one billion people lack shelter fit for human habitation, and most of these are in the cities of the Third World (UN Centre for Human Settlements, 1990: 2). When the new millennium starts, for the first time in history cities will be home for half the people on the planet. New types of urban environment are emerging, and more urgent questions are being asked about the world's cities than at any time since the invention of the industrial city almost 200 years ago. United Nations projections suggest that 'Third World' urban population will grow by more than 700 million persons between 1990 and 2000, and that 80 per cent of the world's population growth in this decade will be in urban areas (UN, 1991). In 1960, eight of the world's ten largest urban areas in terms of population were in the First World. By the year 2000, eight of the ten biggest cities will be in Asia or South America, and they will be bigger than any cities previously known.

Lowe (1992: 130) suggests that between 70 and 95 per cent of new housing in most Third World cities is unauthorised. Sewer networks are expensive to install, and the problems are compounded where buildings are dense and layouts irregular, as they typically are in unauthorised settlements. Poor families without piped water are most likely to drink polluted supplies and hence to fall victim to dysentery and other forms of water-borne diseases. Women and young children are the main victims of these social inequities. Diseases that can be eradicated when there is an adequate water supply and sewerage system today cause huge human suffering in Third-World cities. In Pakistan, water-borne diseases cause up to 40 per cent of illnesses and 60 per cent of infants' illnesses. Diarrhoea is the biggest killer of infants in Third-World cities.

How well has town planning addressed problems such as these? Typically the town-planning systems of less developed countries were fashioned to address the problems experienced in the home country of a colonial power; for example, the Indian Town Planning Act of 1915 was based on the British 1909 Housing, Town Planning (Etc.) Act. From such traditions came a mode of town planning that was essentially local, physical, restrictive and negative, aiming for orderly suburbanisation. In the post-war period comprehensive development plans were produced by foreign consultants. They were typically static, finite master plans, which implicitly presumed a strong state capable of delivering implementation via public funding and effective control measures; such assumptions are not realistic today.

To take just one example; the city of Faisalabad is Pakistan's third largest city. It is growing rapidly as its textile industries expand within the new international division of labour. A Structure Plan was produced in 1986 which calculated that 14,000 plots per annum had to be developed to meet the housing needs of the growing population. In practice, public agencies have found it impossible to develop, as they have to purchase land at market rates and do not have the resources to do so. Meanwhile only 7,000 plots were developed in approved schemes by the private sector between 1986 and 1993, with no schemes being approved by the Faisalabad Development Authority in 1992 and 1993 (Siddiq, 1994: 6). In some of these approved private schemes planned infrastructure has only been partially provided, whilst in others the quality of works was so poor that the infrastructure deteriorated immediately, with problems of sewage disposal and water supply very common (Siddiq, 1994: 4). Meanwhile the poor often cannot afford the costs of housing in these approved private schemes, or find the location of the schemes (typically away from the built-up areas) too remote from employment opportunities and inadequately served by transport (Siddiq, 1994: 59-60).

This form of town planning lacks credibility. In contrast, international donor agencies are backing *ad hoc*, innovative development projects emphasising private-sector initiatives, community involvement and cost-effective use of limited resources. For example, the Orangi Pilot Project in the slums of Karachi has been used as a model. Aided by a small amount of core external funding, a community-development approach succeeded in bringing down the cost of sewer provision to affordable levels through use of 'sweat equity'. Through involving people at lane and neighbourhood level in the installation and maintenance of the system, the project was able to bring decent sewerage facilities to 600,000 poor people.

DON'T MENTION THE 'P-WORD': URBAN PLANNING AFTER 'ACTUALLY EXISTING SOCIALISM'

Because of its associations with the previous regimes, 'town planning' has become a despised term in former communist countries. In practice, town planners were subservient to the sectoral plans: Bater (1980: 51), for example, noted:

> the difference in economic and political power exerted by planning officials on the one hand and the managers of enterprises which are part of nationally important ministries on the other.... In the contest

> between ministerial priorities and town planning principles it is almost inevitable that the former takes precedence.

The East European experience between 1945 and 1989 shows the failures of central planning and the need for some plurality in the process of urban development. Perhaps the most compelling failure was in relation to the environment, which became an important source of semi-legitimated criticism even before 1989. A local journalist quipped that 'The air pollution, more than the existence of the Iron Curtain, brought about the revolution in Czechoslovakia' (quoted in Hague, 1990: 20). The legacy of environmental damage is noted in a report on Prague, which observes, 'Further deterioration of the ecological conditions in the city is not possible and in this sense the disturbed environment in Prague is the most serious barrier to its further development' (Turba, 1993: 34). There was also a serious failure in housing provision and in housing renewal. Housing was a drain on state expenditures and consequently there was under-investment in housing *vis-à-vis* industry (Szeleyni, 1983). Kansky (1976: 110) noted that in Czechoslovakia:

> The theoretical notion that in a socialist system every family has a right to have a suitable apartment and that the state should provide it within a reasonable period of time has been largely ignored in the 1950s and modified in the 1960s.

The previous basis of town-planning practice in these countries is obsolete, and new mechanisms and new professional attitudes have to be forged for the future. The static long-term master plans, typically twenty-five to thirty years in time horizons, and backed by lavishly drawn zone plans to develop massive high-rise housing complexes, are dead whatever their technical legal status. Nobody will pay planners to amass volumes of statistics about topics such as 'social facilities in medium sized towns'. Distant local councils will not engage the central planning-research agencies which used to prepare the local plans: indeed many feel they can dispense with plans altogether. Ordinary people will demand a say in the designation of conservation areas, rather than leaving such matters to trained 'scientific experts'. It is not feasible now to assemble a large multi-disciplinary team of professionals to define or solve any problem.

Transition to a market economy requires that the future of town planning has to be radically different from its past, though as Maier (1994: 264) notes:

> It will take some time to establish a workable planning doctrine to compete with and replace the former ideology of centralized, socio-economic planning ... none of the planning models as they have evolved in developed countries with mixed market, post-affluent societies can be passively transferred to the turbulent under-affluent environment of a post-communist country.

Planners now need to understand the dynamics and agencies of property markets. They must plan for more diverse client groups (indeed, they first have to grasp the concept of 'client groups'). Planners will have to become more entrepreneurial, learning to spot opportunities, but also gaining some understanding of development appraisal so that they can begin to evaluate the options open to them. As Hammersley *et al.* (1994) note, the planning control system will have to find ways to cope with 'the pressures of commercialism in a free market economy'.

Democratisation means new attitudes and skills are needed so that the views of the public can become a part of the planning process. The environment must be given a new priority and the focus of planning effort will have to shift from the major capital-intensive projects which aggrandised the state towards the more complex process of managing and renewing the existing built environment. The legacy of cheaply fashioned panel-construction high flats is a timebomb ticking away, obscured for the present by housing shortages and any number of other problems. But already the Czech word for panel houses, '*panelaky*', carries contemptuous connotations, and in the not too distant future these massive estates will pose major challenges in urban management.

Can we discern something of the form town planning might take in the future in this part of the world? The southern Bohemian town of Český Krumlov has a population of 12,000. Since 1992 it has figured on UNESCO's list of 300 architectural sites of world importance, because of the quality of its historic townscape. A paper mill six kilometres south of the town was a major employer under the communists, but also a major polluter of the river and the atmosphere because of its old technology and reliance on brown coal. After 1989 the town council saw new possibilities for the town, and since then has tried to fashion a regeneration based on culture, ecology and public–private partnership. In February 1992 the town council founded the Český Krumlov Development Fund Limited, providing it with over fifty strategically important buildings in the town centre, including two major hotels, a brewery, shops and buildings used as offices. The fund is an independent business, separate from, but owned by the council. Its main aims are 'to manage, rent, lease, form joint-ventures or sell the properties received from the town to the benefit of the town and its citizens, in accordance with guidelines set by the Town Council'. Foundations have been set up linked to the fund. The Social Foundation aims to help people with low incomes and disabilities, and one of its first steps has been to build a hostel for homeless people. The Cultural Foundation will support 'quality cultural activities that are less commercially attractive'. The Sports Foundation is supporting sports teams for young people. The Schools Foundation will 'support ecological education of people, teaching of arts, with special attention to national heritage protection and the restoration of art works'. Industrial pollution has been tackled. The council has also given the stigmatised gypsy minority jobs keeping the streets clean.

A DEREGULATED FUTURE? LESSONS FROM BRITAIN

We have seen that less-developed countries and the 'under-affluent post-communist' countries have aspired to deregulated, market-led economies, or had them imposed via the 'structural adjustment mechanisms' required by the world's financial and development agencies. For a decade the UK planning system was subjected to the major ideological current that swept the metropolitan world in the 1980s – *laissez-faire* capitalism. The British government, proud torchbearers for the cause of 'rolling back the state', undertook a sustained and committed attempt to change the planning system. If the communist states demonstrate the failures of central planning, the results of Britain's deregulation demonstrate why market economies need town planning. The closer the government came to the abolition of town planning the more disastrous were the consequences.

The changes made to the British planning system as a consequence of Thatcherite ideology have been widely documented (Ambrose, 1986; Griffiths, 1990; Thornley, 1991; Ward, 1994). What has been less noted is the extent to which the Thatcherite understanding of town planning was shaped by a very specific geographical context, namely the suburbia and small towns of southern England. Of course this was precisely its political constituency, but the crucial point is that the political cleavage in 1980s Britain matched an economic divide in the strengths of property markets and the dynamic of the urban-development process. This is relevant both to explaining the collapse of the Thatcherite project and also to understanding the relation between town planning and recession.

Not surprisingly, faith in the untrammelled market was strongest where the market itself was strongest. The idea that deregulation was a sufficient answer to the problems of the inner cities in the North, Scotland or Wales was flawed because of the weakness of markets there. In these places, where the mid-1980s 'boom' was slow to arrive then quickly waned, public agencies played a crucial role in prompting and funding development, with the Scottish Development Agency (SDA) and the development corporations in Scotland's new towns being arguably the most successful examples. This stood in contradiction to purist Thatcherism, and eventually the SDA was broken up into less effective Local Enterprise Councils and the New Town Corporations were wound up. However, it is important to recognise that the success of these public-sector bodies lay in their ability to understand and work within the market. It was not just their investment capabilities that were significant but the skills and attitudes of the staff, and their ability to command the confidence not only of the private sector but also of the public sector. In Scotland in particular there was no imposition of the Thatcherite flagship urban-policy initiative, the Urban Development Corporation. Instead, a model developed of public – private – community partnership, which was later replicated in English urban regeneration programmes.

Deregulation of planning encountered different problems in the Thatcherite heartlands. Booming house prices spurred developers, but protection of property values also mobilised people into resisting development. 'NIMBY-ism' ensured continuity in green-belt policy, for example, and undermined enthusiasm for a *laissez-faire* approach to development. The involvement of town planners in issues of design quality further illustrates the U-turn that was performed by Britain's Conservative governments in relation to the control of development. In 1980 planners were told that they 'should not impose their tastes on developers simply because they believe them to be superior' (DOE, 1980: para. 19). By the 1990s the tone was significantly different, as recognition dawned that the quality of the urban environment was important to the quality of life and an influence on the competitiveness of locations, and that the market could be assisted to deliver such quality. In 1994 the Department of the Environment (DOE) launched its 'Quality in Town and Country Initiative', which conceded that there could be a tension between long-term quality and short-term profit.

In England, strategic planning was discarded more easily than planning control, notably by the abolition of the metropolitan counties and the Greater London Council (GLC) in 1986. As Ward (1994: 258) notes in reviewing the period from 1974 to 1994, 'Overall the consistent political antipathy to coherent strategic planning remains the strongest impression of these years.' The results were entirely predictable as the recession followed the boom that had been fuelled by house-price inflation and consumer credit. The out-of-town or edge-of-city shopping malls approved in the mid-1980s combined with the recession to create a crisis in the town centres in the early 1990s. Consequently, the Secretary of State for the Environment introduced new guidelines in 1994 that were designed to stop new applications for out-of-town retail parks. A decade of playing the market saw a reassertion by government of the need for plans.

The lack of strategic planning for London ensured that competition between the City and Docklands produced an overprovision of offices, so that there were 34.2 million square feet of empty offices in prime central-London locations in 1992, according to property advisers Debenham Tewson and Chinnocks (Dovkants and Bar-Hillel, 1992). The empty Canary Wharf, haunted by the ghosts of Olympia and York, its bankrupt developers, stood as the outstanding monument to Thatcherite urban alchemy. A key reason for the problem of Canary Wharf was the lack of adequate transport links, a symptom of the wider transport crisis in London, and a clear illustration of the need for integrated planning of land use and transport.

The British experience also shows the international green agenda forcing its way into town planning despite the antipathy of the Thatcher government to the ideas and ethos of environmentalism. Undoubtedly, key influences came from outside Britain, notably the report of the United Nations World Commission on Environment and Development (WCED, 1987). This was followed by the European Commission's Green Paper on the Urban

Environment (CEC, 1990). These documents set the keynotes for the British White Paper, *This Common Inheritance* (DOE, 1990). The growing international concern with sustainable development coincided with the NIMBY reaction against development pressures to push the Conservative government into 'greening' its planning policies. Significantly, this greening has survived the recession, and environmental concerns have been a growing feature of the work of British town planners. The British experience therefore suggests that sustainability will continue to be an important influence on town planning into the twenty-first century, despite the problems of the concept as discussed in Chapter 1.

In terms of the urban poor, the Thatcherite equation of property development with urban regeneration and reliance on the 'trickle down' of benefits was a flawed strategy, even in the boom experienced in southern Britain in the 1980s. In recession it is an irrelevance. A key need for town planning in the future is to address the needs of those places and people whose market power is weak. This requires compensatory public investment, and strategies that are deliberately based on a partnership with communities, mobilising human resources and empowering people. British planners have become more aware of the needs of women, ethnic minorities and disabled people than was the case in the early 1980s. 'New Urban Left' local authorities, most notably the GLC, led the way, but increasing pressure from disadvantaged groups themselves has also been important. Last, but not least, the Royal Town Planning Institute (RTPI), has made teaching about equal opportunities a mandatory requirement on courses seeking its professional accreditation, and has promoted and commissioned work on good practice in this field. Of course the advance has been uneven, as Krishnarayan and Thomas (1993) show in relation to planning authorities' awareness of ethnic-minority needs. Nevertheless the signs are that an appreciation of pluralism will increasingly infuse the delivery of the planning service.

Despite the debacles and U-turns, the Thatcherite project undoubtedly achieved significant changes in British town planning. Planning became an enabling function which worked in partnership not just with other public-sector agencies but with the private sector and voluntary organisations. The skills and attitudes of many planners changed significantly over this period. They became more entrepreneurial in outlook, more aware of the diversity of needs to be met and of the global environmental agenda, better at negotiation and implementation. It is also significant that planners began to work in many roles outside the statutory town and country planning system, in transport, for housing associations, in countryside management, economic development, etc. The bulk of Britain's planners are still involved in local-authority planning departments, and of course there are still planners who operate in a bureaucratic mode. As noted above, there have also been important continuities in aspects of planners' ideology and ethos. Nevertheless, town planning in 1990s Britain is significantly different from the practice that characterised 'classical town planning'. It is a form of planning practice suited to the context set by recession, resources and pluralism. If we peer into the future from the British experience we see the need for strategic planning and for public agencies able to work effectively with the private sector. Yet planning will also need to become more green and more decentralised, and pluralism is encompassing the planning profession itself as planners take on a wider range of roles.

DIVERSE WORLDS, ONE FUTURE?

This chapter has argued that recovery from recession will be slow and the booms will be weaker than in the past. There will be increasing concern about how to generate resources to fund development (and hence increased partnerships between public, private and voluntary sectors). Issues of sustainability and conflicts about the meaning of sustainable development will become increasingly important. Likewise, town planning will have to come to terms with the increased plurality of societies; the differences between rich and poor, males and females, young and old, and different ethnic groupings will make planning a more complex and negotiative process demanding more sophisticated 'people skills', and a planning profession more representative of the whole society.

The review of the international context and experience of town planning suggests some broad conclusions. Despite the very different rates and forms of urbanisation in different countries, there appear to be some similarities in the trajectory taking town planning into the twenty-first century. These involve some continuities with past planning ideas, but also some important changes. In particular:

1. *Town planning will not be perfect but it is a better option than simply relying on unrestrained market forces.*

Since 1979 the British government has made no secret of the fact that it sees market forces as a better basis for organising society than state planning. Yet we still have a planning system, and in line with other European Union countries we have moved towards a 'plan-led system' of development. Similarly, after the first flush of enthusiasm for the free market, there is now a recognition in central and eastern Europe of the need for plans and effective control systems. In less-developed countries 'classical town planning' is likely to become 'development planning'. The relation between planning and the market can take diverse forms and is likely to depend on circumstances, but planning still appears to be needed.

2. *Town planning will become more broad-based, promotional and flexible.*

The traditional basis of town planning has been undermined by recession, political change or rapid urbanisation. The northern European planning tradition was basically about controlling edge-of-city expansion, and consequently was weak on promoting and enabling development. This tradition was exported to colonies. Where rural–urban migration is rapid and driven by lack of rural opportunity, this form of town planning lacks credibility. Preparation of detailed land-use plans takes a long time, and cannot keep pace with the urbanisation process. Plans become out of date and discredited as a basis for implementation. There is an underlying premise that the city is a problem, not an opportunity. Likewise, in the 'post-communist' countries and the 'post-affluent' developed world, recession means that town planning must be a means of facilitating and enabling development.

3. *Town planning needs to provide a vision and a strategy.*

One of the strengths of the town-planning tradition has been its insistence on the need for a vision of the kind of places in which we want to live. There are weaknesses in this too – at times it has been the overarching vision of one man, at other times the vision has been lost beneath a morass of bureaucratic procedures. The need for imaginative thinking about cities is greater than ever. A vision of new possibilities is vital to the success of ventures such as the Orangi Pilot Project or the Český Krumlov regeneration. Though projects such as these extend beyond the boundaries of traditional town planning, one of the strengths of town-planning thought has been the idea of comprehensiveness, which includes the recognition that physical, economic and social factors are interconnected. Planners have also recognised the importance of translating the vision into a strategy. These understandings must underpin town planning in the next century.

4. *Planning needs to be based on principles of sustainable development and partnership.*

Some of the key ideas about sustainable environments have always been part of the town-planning movement. However, the Earth Summit and Local Agenda 21 have given a new significance to these ideas, and have linked environmental preservation with issues of social equity and participation. Concerns for sustainability are not the exclusive preserve of town planners, and town planners by themselves are unlikely to be able to implement such policies. Increasingly effective town planning will mean partnership with the community and with the private sector. The old dualism of professional and community will become less relevant in a world of flat hierarchies, flexible labour markets and mounting concern about urban and environmental issues. Of course there are contradictions and conflicts in this, but these emphasise the need for planners to work positively at trying to build coalitions for positive and sustainable change.

The globalisation of markets and the international nature of the environmental problems that we face are pushing town planning towards a common global agenda. Public authorities are short of money and there is a strong presumption that the private sector will play a key role in future development. Yet it is also clear that free markets are not a satisfactory basis for managing urban change. In the future, planners everywhere will have to operate in that blurred area where state and market and voluntary agencies overlap. The prescription is therefore that town planning will need to be client-based, negotiative and committed to making limited resources go a long way without damaging the environment. The process will itself be one of the ends. Planners will need new knowledge about ecological processes, a better understanding of markets and the economics of development, new awareness of the diversity of social groups and their conflicting needs. Above all we will need 'planners with attitude', a professional commitment to an ethic of public service that is enterprising and innovative, whilst holding firm to the need for a long-term safeguarding of natural resources and to improving the opportunities of those who lack power in the market.

'Classical town planning', as defined in the opening chapter of this book may indeed be dead, but the professionalism and reform traditions of British town planning still provide a valuable platform from which to view the future. The management of urban change will be a major issue as we enter the next millennium. Those prepared to grapple with it will define the future of town planning.

9

SOCIETY AND SUSTAINABILITY

The context of change for planning

Andrew Blowers

A MOMENT OF TRANSITION OR TRANSFORMATION?

The environment has become established as a political issue of global significance. This is evident in the scale of the environmental problems and the prospect of global catastrophe they present, and in the political response they have engendered. It is conceivable that we have reached a defining moment in the relationship between the environment and society whereby changes in economic, political and cultural dimensions are occurring both in response to environmental change and in order to prevent changes that may threaten survival. Whether it is a moment of transition whereby an accommodation with environmental constraints is secured through adaptation, or, alternatively, a moment of transformation leading to fundamental social changes is a matter on which this chapter will speculate. What is clear is that the notion of *sustainable development* has been appropriated both by those who believe changes can be achieved without impeding the continued march of economic progress and by those who consider that we are doomed unless we abandon the destructive pursuit of modern forms of economic growth. This debate over sustainable development is the central theme of this chapter.

So far the debate has been engaged largely at a rhetorical level and the issue of what changes will or should occur and how they are to be managed has been largely sidelined. The role of planning in the process of change has been almost totally ignored, an oversight which this chapter seeks to redress.

'Planning' in this context is both narrowly and broadly defined. On the one hand, it is that process, described in Chapters 4 and 5, that relates to the legal and administrative function of town and country planning which regulates the use of land through procedures of development plans and development control. This is the domain of 'professional' planning, a government activity which, in principle, is intended to ensure that the public interest is taken into account in the distribution and allocation of land to particular uses, though whether planning, in practice, functions to support private interests is a matter of debate as Chapter 5 has demonstrated. For the purposes of this chapter, planning as a legitimate arm of government is one of a number of activities that can help to secure sustainable development through setting and implementing targets and encouraging patterns of land use that help to reduce resource consumption and pollution.

At a broader level, planning is also defined as a means of long-term environmental management. This incorporates both purpose and process. The process of environmental planning as a holistic, integrative and strategic form of intervention is set out in Chapter 3. Its more purposive role, a concern for environmental protection and greater social equality (aspects of which were discussed in Chapters 3 to 7) both revives some of the original objectives of planning and casts them in the contemporary light of sustainable development. In this context, planners are a group working alongside other interests in civil society (scientists, non-governmental organisations (NGOs) and, to some extent, business) whose purpose is to promote the needs of the environment as an integral part of social and economic development.

In what follows, both definitions are used and will be identified where appropriate. The contrasting definitions relate to a continuing debate about planning which has been an underlying theme throughout this book. Planning both in its narrow and in its broad sense is relevant to the process of social change that is necessary if sustainability is to be secured.

The chapter is in four sections. The first focuses on the nature of the contemporary environmental problem (or, as some prefer, crisis) and the emergence of the idea of sustainable development as the means to solve it. Alternative perspectives on the approach to solutions are discussed in the second section. This focuses on the idea of continuity or transition conveyed by ecological modernisation as opposed to the idea of fundamental change or transformation presented by advocates of the 'risk society' thesis. The possibilities for change are analysed in the third section which identifies the role of social movements in introducing and influencing a new environmental politics. Finally, in the fourth section, the prospects for change are considered and a key role for planning is envisaged.

THE ENVIRONMENT – PROBLEM OR CRISIS?

Characteristics of environmental problems

Although public concern about environmental issues waxes and wanes, the threat to survival posed by the depletion of resources and the pollution of ecosystems has become a matter for local, national and global action at the political level. Different environmental issues hold the stage from time to time. During the 1980s, and especially after Chernobyl, the threats of nuclear accident and proliferation, were a pervading source of anxiety. The ending of the Cold War has reduced fears of nuclear conflict (though the possibilities of a major accident or of conflicts arising through proliferation persist). Other problems have re-emerged, such as the pollution of the oceans or the so-called 'desertification' of drylands. Other 'new' problems, invisible, yet defined by scientists, have also engaged attention, the most prominent being the depletion of the ozone layer, the loss of biodiversity and the global warming brought about by the enhanced greenhouse effect.

We appear to have entered a period of global ecological risk brought about by environmental problems which, in certain respects, are distinctive from those of previous times. First, these problems are always *anthropogenic in origin*. It is the impact of human activities on the natural environment that results in resource depletion and environmental degradation. While it is true that some of these activities (such as deforestation or over-exploitation of soil) are nothing new, they, along with the pollution and damage caused by modern industrial processes, threaten, perhaps for the first time, the overall capacity of natural systems to cope with the burdens placed upon them.

Second, these problems are *global in reach*. This may arise from the diffusion of sources of pollution around the globe as a result of the spread of modern agricultural and industrial activities. In addition there is the diffusion of impacts throughout the atmosphere or oceans which may, ultimately, affect the survival of human society on the planet.

However, a third feature is the tendency for impacts or the ability to deal with them to be *socially uneven*. Benefits and burdens are unequally shared. Polluting activities become concentrated in so-called 'pollution havens' or 'peripheral communities'. Poorer regions or countries are unable to defend themselves from the impacts. In the past, social inequality was often associated with environmental inequality on a local scale; now it has a global dimension. In addition, the long-term and often incremental impact of many environmental problems introduces an inter-generational dimension whereby the needs of future generations may be forfeited by the demands of the present.

A *concern for survival* is the fourth feature of contemporary environmental problems. This concern embraces both the loss of resources necessary to sustain life-styles and the deterioration in the environment essential to support life.

A fifth aspect is the importance of *experts* both in identifying and providing solutions for environmental problems. There is now a culture of expertise which influences the priority given to issues and the political willingness to deal with them. Yet scientific knowledge, apparently so authoritative, is uncertain, contestable and consequently refutable. Without access to expertise it becomes impossible to challenge the assumptions and implications portrayed by experts. Whether or not planners may be regarded as experts has been debated in earlier chapters (4 and 5). Whatever one's view of the matter there is a need for an interdisciplinary perspective that planners can bring to bear on the analysis of environmental problems which, in itself, is perhaps a form of expertise.

Experts constitute one of several interest groups concerned with environmental problems. *Conflicts between interests* form a sixth characteristic. Other interests include business corporations, environmental movements, nation states, inter-governmental institutions like the World Bank, the International Trade Organisation, the IMF or groupings of countries such as the EU. The conflicts between interests are engaged both spatially (between North and South especially) and over time (inter-generational). In political terms they are conflicts over the relative priorities to be accorded to economic growth or environmental protection.

The conflict between economy and environment possesses an *ethical dimension*, the seventh feature of environmental problems. This conveys the values that shape attitudes to the environment and the ideologies that influence interests and policies. A wide range of values is involved including the concern for the intrinsic value of nature, the rights of individuals and future generations, the nature of participation in decision making, the importance of social equality and so on.

Sustainable development

These seven characteristics indicate that environmental problems cannot simply be considered as physical phenomena; they have social aspects too. Yet the debate about sustainable development has tended to focus on the 'sustainability' aspect, on the processes of physical change and their implications. Sustainable development as a concept invites us to recognise both the physical (sustainable) and the social (development) aspects and the interaction between them. But sustainable development is a multi-dimensional concept; it is, at once, a scientific principle and a political goal, a social practice and a moral guideline.

Hitherto, much effort has been expended in seeking definitions of sustainable development, much less on trying to operationalise the concept in terms of what a sustainable society might be like and what is necessary to achieve it. The problem with definitions is that they tend often to be ideological, to reflect particular constructions of reality. Given the various, and often self-serving, definitions of sustainable development, it is small wonder that the debate has remained largely at a rhetorical level. But the different definitions of sustainable development are at the heart of a major discourse about the nature of modern development which has ideological and political dimensions. This discourse reflects a fundamental conflict, already referred to, between those who believe that sustainable development can be achieved without seriously impeding contemporary economic processes and those who argue that sustainability is impossible unless there is a quite different form of development.

Broadly, the conflict can be defined in terms of contrasting perspectives on sustainability, termed respectively 'weak' and 'strong' sustainability. Although they are idealisations they reflect fundamentally different attitudes to the problem of achieving sustainable development. On the one hand, so-called 'weak sustainability' is anthropocentric, emphasising human survival and the compatibility of economic development and environmental conservation through the substitutability of resources and greater emphasis on the reduction of polluting processes. It is a reformist approach, insisting that sustainability can be achieved through greater emphasis on environmental constraints. Moreover, such a form of development is thought to be consistent with modern liberal economic and democratic systems.

By contrast, ideas of 'strong' sustainablity are very exacting. They emphasise the significance of all forms of life, the need to protect and conserve resources and believe the conflicts between economy and environment are irreconcilable. Representative of this stance is the 'deep ecology' of Arne Naess with its seven principles, as follows – rejection of the man-in-environment image, biospherical egalitarianism, diversity and symbiosis, anti-class posture, fight against pollution and resource depletion, complexity, local autonomy and decentralisation (Naess, in Dobson, 1991, pp. 42–7). A strong version of sustainability considers present trends as inevitably destructive and that they will, sooner or later, threaten survival. Consequently, economic systems must be transformed to secure sustainability and the political system must also be changed to enable the long-term interests of environmental security to flourish.

APPROACHES TO SUSTAINABLE DEVELOPMENT

These different perspectives are reflected in contrasting approaches to the problem of securing sustainable development. The first approach, often called 'ecological modernisation' is conservative in the sense that it seeks conservation through continuity of the economic system. The second may be termed the 'risk society' analysis, since it draws on the work of Ulrich Beck who propounded the ideas in a series of texts (Beck, 1992, 1995, Beck *et al.*, 1994). This approach argues that contemporary systems are sowing the seeds of their ultimate destruction, that sustainable development means the rejection of modern high-risk technologies and the adoption of alternative (albeit unspecified) modes of production together with greater participation and democracy. Planning has a part to play in each approach, though under very different definitions and conditions in each case.

Ecological modernisation

The most striking – and comforting – aspect of ecological modernisation theory (EMT) is its contention that the environment and the economy are not in conflict; rather that environmental protection can only be secured through economic prosperity. This viewpoint has clearly influenced many policy statements including, for example, the UK's position statement on sustainable development:

> Sustainable development does not mean having less economic development: on the contrary, a healthy economy is better able to generate the resources to meet people's needs, and new investment and environmental improvement often go hand in hand.
>
> (HMSO, 1994a, p. 7)

EMT is, therefore, on the whole an optimistic approach. It speaks in the language of consensus, it appears to be both rational and realistic and it fits neatly into the liberal economy and conservative value systems which pervade the contemporary western world.

Following Mol (1995), there are six basic premises of EMT. First is the prominence it accords to ecological criteria and environmental needs in production and consumption processes. Second, it emphasises the ability of science and technology to 'refine' production, to achieve greater environmental performance through greater economic efficiency. Third, it promotes the market as the most effective means of securing the flexibility and responsiveness necessary for ecological adaptation. The state's function is perceived as facilitating favourable conditions in which the market can operate and as providing a regulatory framework and standards of environmental performance. Fourth, environmental movements are seen as functional to ecological modernisation in so far as they can be incorporated into the decision-making process. Indeed, EMT has a neo-corporatist flavour with its emphasis on partnership, participation and notions of a stakeholder society. Fifth, multi-national companies, the progenitors of the global economy, are regarded as the leading agents of change. And, finally, opposition to ecological modernisation is dismissed as impractical, lacking in support and, consequently, of marginal significance.

Essentially, EMT confirms a business-as-usual approach to sustainable development. It goes with the grain of present-day consensus based on an enabling state and a capitalist market economy. It can appeal both to the New Right, which echoes its main tenets, and to the 'modern' New Labour Party with its positioning in favour of a 'stakeholder' society.

In many respects, planning, particularly in its 'professional' definition as currently conceptualised both by its practitioners and by government, fits very neatly into EMT. Planning can be seen as a necessary part of the state's regulatory apparatus, largely devoid of ideology, functioning to facilitate market processes. This perceived role is evident in a whole swathe of documents from *This Common Inheritance* (DOE, 1990), through planning policy guidance, to structure and local plans. Planning is seen as a means to *guide* development, to *avoid* conflict, to *facilitate* growth and to *protect* resources and amenity. There is an emphasis on balancing demand and resources, on securing community support through consultation and on encouraging partnerships between private- and public-sector interests. Thus, planning serves to legitimate the capitalist economy by helping it to function efficiently. It is a limited, pragmatic but practical view that does not challenge the prevailing systems in any way.

EMT is a descriptive theory that makes a virtue of the inevitable, celebrating contemporary capitalism but making it environmentally benign. It is also a restricted vision in that it neither addresses some basic social issues, notably inequality, nor offers solutions for certain global environmental problems that are likely to be exacerbated by continuing exploitation of resources and pollution. The threats posed by modern technologies and their social aspects are the focus of the risk-society thesis.

Risk society

'Risk society' is the term used by Beck to describe the consequences of modern technology, which has produced 'the self-created possibility, hidden at first, of the self-destruction of all life on this earth' (1995, p. 67). His thesis focuses on the social consequences of environmental change. It is a pessimistic thesis, conceived originally when the Cold War threatened the possibility of nuclear exchange and Chernobyl signalled the consequences of a major nuclear accident. Thus, the analysis presents conflicts between ecological needs and economic demands, presumes a tendency towards authoritarian control and predicts catastrophe unless contemporary trends are arrested. In stark contrast to EMT, the risk-society thesis concludes that a social transformation is necessary if we are to ensure survival.

It is possible to interpret the thesis in terms of three basic elements. The first focuses on the risks from modern technologies. These risks are indiscriminate, unpredictable and, in some cases, irreversible. 'More and more, the centre comes to be occupied by threats that are often neither visible nor tangible to the lay public, threats that will not even take their toll in the life-span of the affected individuals' (Beck, 1992, p. 162). They are the result of modern production and consumption patterns. But no individual nor even society, on its own, can control them. The benefits of modern technology are dispersed as are the risks, and since we can consume while at the same time avoiding responsibility for the environmental effects, there is little incentive to desist from production. This would require collective action in the general interest asserting its will over individual and corporate private interests – a situation that has little appeal in the current political climate.

The second element of the thesis concerns the dependency on experts, a point raised earlier in this chapter. Decision making in risk society is heavily influenced by those with knowledge and expertise. They are vested with the control of risk-creating technologies. Since risks can never be altogether eliminated (Chernobyl alone is sufficient evidence for that), the possibilities of hazards occurring must be presented as infinitesimally small, so as to appear almost impossible. Thus the risks of a death from a radioactive release are indicated by such technical criteria as one in a million risk and so on.

There are two problems here. One is that if the improbable should ever occur it could be catastrophic. The other is that the experts themselves are dealing with matters of uncertainty over which there can be, and is, considerable disagreement and also considerable potential for error. But the methods of risk evaluation serve to calculate the possibility of risk, not to prevent it altogether. 'In an age of world wide growth of large-scale technological systems, the least likely event will occur in the long run' (Beck, 1995, p. 1).

Thus experts hold enormous power. Decision makers cannot act without expert advice and the fears of citizens or the claims of environmental movements are thereby excluded. Moreover, the experts themselves are frequently acting as an interest group on behalf of major industries or of governments who depend on such industries for economic performance. This condition leads to an authoritarianism that denies democratic participation. However, uncertainty itself breeds dissent. Experts may come not to be trusted or their advice may be contested. Increasingly, counter-expertise is developed to challenge the conventional wisdom. It is here that environmental movements begin to succeed in defeating the more preposterous proposals of the experts. It is here, too, that planning, itself a profession with claims to expertise, has a potential role in arbitrating between competing claims about the environmental consequences of modern technologies.

The uncertainty and insecurity experienced at the individual level in society more broadly constitutes the third element of the thesis. This condition is called 'individualisation' by Beck. It comprises three aspects. One is the economic insecurity experienced as a result of industrial restructuring, consequent unemployment and loss of long-term job security. This is compounded by the progressive withdrawal or reduction of the welfare-state support systems including health, education and retirement benefits. And the third, partly connected to the first two, is the personal dislocation now widely experienced and manifested in marital breakdown, fear of crime, poverty and social polarisation. Traditional patterns of social and personal integration such as marriage, family, kinship, neighbourhood, religion and community, along with more modern ones like trade unions or political parties, seem to have broken down, leaving a society that has become more fragmented and individualised. This may lead to a social condition of fatalism, of a search for transitory satisfactions. Or, possibly, it may open up, for some people, the possibilities for major shifts in attitudes and values.

Interpretation of these social changes and their implications for the environment (among other things) is another area that planners must consider if they are to make sense of the relationship between social change and the environment.

Like EMT, the risk-society thesis must be treated with some reservations. Apart from its specific western focus, it, too, whilst acknowledging the inequalities of power in risk societies, tends to treat society at a general level, thereby ignoring the unevenness of impact of risky technologies on communities and social groups. By concentrating on contemporary technological risks it also neglects those long-standing risks (of famine, war or disease) in past epochs and in poorer societies which were (and are) a quotidien condition. Unlike EMT, however, risk society is an analysis (even a polemic) which exposes the social consequences of technological change, not a proposition on how change is to be managed. To that extent it is utopian and idealistic, offering few solutions beyond an obligatory warning that we must desist in favour of the vague idea of a new Enlightenment.

THE PROSPECTS FOR CHANGE

The role of planning

The two approaches reach entirely opposite conclusions and are, therefore, apparently irreconcilable. On the one hand, EMT suggests that environmental degradation can be minimised through the refinement of production under the aegis of an enabling state. Progress is both technically possible and politically feasible without injuring the present institutional arrangements and social structures. On the other hand, risk society concludes that modern production systems are inherently dangerous and a potent threat to survival. Therefore they must be fundamentally changed along with the institutions and political and economic processes that support them. Since the alternative is unclear, the thesis can easily be dismissed as unrealistic and unacceptable.

Planning is relevant to both positions, though in fundamentally different ways. This reveals the two definitions of planning outlined at the beginning of this chapter. Under EMT, planning is a responsive process. It can be envisaged as part of the state's regulatory apparatus ensuring that environmental dangers are recognised and standards are met, arbitrating conflicts between economic and ecological demands on natural resources and mediating between public and private interests. With risk society, the role of planning is more overtly ideological, an interventionist process defending the environment against the depredations of modern technologies. Planning, in this sense, by its analysis of social change may find itself aligning with those who challenge the prevailing values embedded in modern technological and economic systems. Thus, these two approaches reveal the underlying debate within planning between those who envisage a limited, professional and non-interventionist role and those who advocate a more zealous, radical and participative role. In the one view planning is incorporated within the state system, pragmatic and neutral; in the other it is outside it, aloof and partisan.

The two roles are not wholly incompatible. The evidence from conflicts over environmental issues suggests elements of both EMT and risk society in the potential role for planners. The evidence offered here is, admittedly, taken from disputes over Locally Unwanted Land Uses (LULUs), those activities that are specifically located (such as power stations, waste facilities, or road, airport and other transport projects) and that tend to be resisted because they bring, overall, more costs than benefits to local communities. Examples are nuclear projects or hazardous industries, and there are numerous examples of the course and outcomes of such conflicts that offer some interesting pointers to the prospects for change.

Environmental conflicts

Disputes over LULUs demonstrate novel features that may be said to constitute a new politics. They are *coalition building*, frequently bringing together coalitions of protest that cut across conventional social and political divides. This *cross-cutting* may embrace Right and Left, cross class boundaries and include supporters of the activity (as long as it is elsewhere) as well as opponents (against it wherever it is located). Thus, in terms of values, these coalitions may include conservative ideologies concerned with defence of property and conservation of the environment as amenity which relate to a utilitarian position. But they also include more radical positions concerned with protection of the commons and the rights of future generations, adopting an ecocentric and communitarian stance. In combination, these divergent interests serve to resurrect traditional integrating systems of community. In opposing the elitist decision-making systems of modern technology, they advocate greater openness and participation. Moreover, they often act in combination with other communities or with environmental movements engaged in national and international action.

They are often up against alliances which, in the economic sphere, bring together hitherto opposed interests of capital and labour. These interests act together in defence of jobs, of wealth and of economic prosperity. Thus, environmental disputes of this type are frequently engagements between opponents operating defensively and drawing together groups which cut right across the sectoral divides that have been created by modern society. Interestingly, planners may find themselves on either side, some acting in defence of local territory and some commissioned by industry to support the case for development.

It would be unwise to generalise too much on the basis of these rather specific conflicts. They cannot be applied directly to those conflicts over more dispersed activities – such as some modern food-production systems or the use of private transport – that create widespread environmental damage but that bring obvious benefits. In addition, disputes over LULUs tend to be ephemeral, bringing together temporary alliances, united over one issue and disengaging once the issue is settled. Nevertheless, it is possible to discern a new politics of the environment emerging in such disputes, fragmented, temporary and insubstantial as it might seem.

Implications for the two approaches

This new politics emanating from the emergence of environmental movements operates both at the level of values and policies. It recognises the possibilities for change and can be applied both to EMT and to risk-society positions. In both approaches the role of science and technology is seen as important. In EMT it is vital to the efficient environmental performance of industry and enables a response to the demands of environmental groups for greater safety and less pollution. In risk society, science and technology are the cause of risk and the means by which it is perpetuated; a view which supports those environmentalists who remain implacably opposed to high-risk technologies. In both approaches there is a stress on the international dimension of modern technology, EMT stressing the ecological innovation it brings, risk society underlining the globalisation of risk that results.

There are also some specific conclusions that can be drawn for each alternative. In terms of EMT, these conflicts can be seen as limited instances, helping to resolve the competing needs of economy and environment without significantly impeding the march of progress. In terms of outcomes, they may lead either to displacement of an activity elsewhere, to its deferral or possibly to its abandonment. In each case they serve to ensure that environmental and economic needs are ultimately met. But they also reinforce existing patterns of location, of power and of values. Unwanted activities frequently land up in 'peripheral communities' (Blowers and Leroy, 1994), those communities that already have a degraded environment or a concentration of risky technologies, that are relatively remote, monocultural in their economic dependence on single or a limited range of polluting industries and hence powerless to resist.

Decision making may become ostensibly more open, encouraging limited participation, but the power structure remains essentially elitist as an accommodation is found between the needs of business and the resistance of organised communities of protest. Protest groups may disappear and some environmental movements may become increasingly incorporated or co-opted into the decision-making process. Above all, the fundamental objectives of the state and of business are not perverted, and prevailing values of economic growth and progress are not impeded.

Environmental conflicts also offer some evidence to support the risk-society thesis. They demonstrate the increasing leverage of environmental movements on prevailing power structures. The success of protest movements may not be merely tactical, collisions in one place leading to retrenchment and reappearance in another form or another place. In terms of specific projects, areas of policy and the wider plain of changing values, the growing influence of environmental movements has yielded palpable returns. For instance, policies for managing nuclear wastes have been slowed down, become more measured and far more stringent as a result of the serious opposition mounted against projects over the last two decades. Dumping in the oceans and the export of hazardous wastes have been largely curtailed. Environmental concerns are now routinely articulated in a wide range of policy pronouncements. Whole regimes of policy from local through national to global have been created in recent years to deal with environmental problems. Of course, these developments often also coincide with economic needs, but the impact of environmental concerns has played a significant role in challenging and changing policies.

Moreover, environmental conflicts may indicate changes in political participation. Environmental movements, as their influence grows, become major actors potentially able to prise open the decision-making process, ensuring a greater openness and accessibility to decision makers. At the same time they are able to retain their independence and integrity while participating in policy making. Co-operation does not necessarily mean co-option. The role of environmental movements in the risk society is not to support prevailing values but to change them.

POLITICS, PLANNING AND SUSTAINABLE DEVELOPMENT

Both ecological modernisation and the risk-society thesis direct attention, in their different ways, to the necessity for sustainable development. For EMT, the environment provides an opportunity for refining industrial processes to minimise degradation. In so doing, it is argued, business becomes both greener and more efficient. For risk society, the environment imposes a constraint on modern technology to the extent that sustainability can only be assured by alternative technologies and compatible forms of social development. In the case of EMT the direction is fairly clear though the consequences may be awful; with risk society the way forward is opaque and the consequences vague and indeterminate.

A remaining question, or rather speculation, then, is what direction change is likely to take and what part planning might play in the process. At first sight, it may seem that continuing development along present lines, albeit modified by some ecological modernisation, is the most likely, indeed inevitable, outcome. The process of economic liberalisation has proceeded unabated and, world-wide, the capitalist economy has reached its apotheosis. Within the nation state the processes of privatisation and deregulation have provided a decisive shift to the private sector. Local-government powers have been curtailed and their finances reduced. Increasing centralisation at national level has been offset both by supra-national functions and by the proliferation of non-elected quangos. Decision making has become less democratic, more elitist. Political parties, in recent years, have become less trenchantly ideological (or at least come closer together on a range of economic and welfare issues) and certainly have ceased to claim the continuing loyalty of a majority of the population. There has been, in Rhodes's words, a 'hollowing out of the state' (1994).

All of this has, on the face of it, enhanced the power of business and is therefore propitious for ecological modernisation. Much power has been removed from obvious democratic accountability, decision making has often become closed and remarkably difficult to penetrate and protests are blocked.

On the other hand, the retreat of the state, the loss of local functions and the decline of political parties has opened up political space which has been increasingly occupied by political social movements including environmental movements. Thus, within a broadly democratic context, there is scope for challenge to be mounted and, using various tactics of campaigning, lobbying, use of media and direct action, environmental movements can seize opportunities to influence policies and to change values.

The prospects for the future rest on the outcome of inevitable present tensions. There is a *common interest* in environmental conservation in the long term. It is necessary for business and for government. But there is a contrary common interest in maintaining the existing situation in the short run. It offers the prospect of jobs, profits and growth. EMT offers a reconciliation of the two, a concern for the long-run environmental need while not impairing the continued economic growth in the short run. Risk society pronounces against such complacency.

The vested and mutual interests in economic growth pursued by business and the nation state seem able to deploy overwhelming power. They are supported by a host of other like-minded interests – the military, civil servants, regulatory bodies and advisory groups – all with a stake in the status quo to a greater or lesser extent. They are practised in the arts of tactical manoeuvre to achieve strategic objectives. Thus a few concessions here, a little greening there gilds the lily of sustainable development. The rhetoric of sustainable development is now so axiomatic that the illusion of change is a substitute for action. The political consensus, stripped of this rhetoric, lays bare the fact that economic interests are still way ahead of any environmental concerns in the minds of those with responsibility for decisions.

Perhaps the best (maybe the only) hope for the environment lies in the potential emergence of a *countervailing power* to challenge the consensus. In part this power rests with scientists, working often in global networks, who have revealed the extent and potential damage arising from global environmental change. Their conclusions are difficult to deny. In part it arises from commercial changes and new technologies or from re-emphasis on more sustainable practices which offer solutions to environmental problems. And, in part, it stems from the development of the environmental movement whose strength grows as it creates a constituency of protest, builds coalitions at all levels and develops as a truly global force. The use of counter-expertise provides credibility, and incorporation in decision making gives legitimacy.

Where does planning stand in all this? At the present time, in Britain at least, planning is largely hitched to the cause of ecological modernisation. This is an important but limiting role, reflecting the narrow definition of planning presented at the beginning of this chapter. As part of the system of government, planning has adapted over recent decades to a much greater emphasis on the market, with its concepts of entrepreneurship, deregulation and partnership. Planning has remained focused on the problem of land-use allocation, though increasingly with an emphasis on achieving objectives of sustainable development through such ideas as reduction of traffic movement, avoidance of pollution, conservation of resources and so on. But, as Chapter 4 has shown, over these years, too, planning has more or less entirely renounced its pretensions to be a force for social change. It is remarkable how much planning has conformed to the new political dispensation, how little it has emphasised the social purposes which were its original foundation, a point given empirical emphasis in Chapter 6.

In its neutered state it would seem planning has relatively little to contribute to the solution of environmental problems beyond adding to the rhetoric of sustainable development, but from a viewpoint of ecological modernisation. In terms of a risk-society approach planning has, in truth, very little to say. But, if the predictions of scientists are to be believed, ecological modernisation will be a palliative, not a solution. The risk-society thesis may be overly pessimistic, especially in its social assumptions, but the prospects of environmental catastrophe have a ring of possibility unless precautionary action is taken.

In the context of the probability of long-term environmental deterioration (and the possibility of annihilation), planning, defined in its broadest sense, could have a vital role to play both at the level of values and at the level of policy making. First, it needs a vision, a rededication of its social purpose in a contemporary context. This vision would have two components. One would be an emphasis on environmental sustainability through the conservation of resources and the prevention of pollution – an assertion of the environmental (not solely land-use) basis for planning. Of course, this broader concern has become evident both in government statements through PPGs and in structure and local plans. Some of these have made sustainable development their *leitmotif*, including targets and indicators to measure progress.

The other would be a concern for social equality. This is overlooked, partly, no doubt, because it is politically contentious. Greater equality is a necessary condition for the achievement of sustainability since it removes the basis for conflict and facilitates co-operation. Put simply, the poor and disadvantaged will see no point in acting sustainably if the rich and powerful do nothing to reduce their own excessive and disproportionate consumption of resources and production of pollution. But social equality is not a sufficient condition since both poor and rich alike are capable of the wasteful and inefficient consumption of resources. This suggests that new forms of production, new life-styles and new patterns of social integration will be necessary. Precisely what forms these will take cannot be envisaged. Planning can assist the process of change by setting out the physical (including land-use) constraints. After all, planning has, in the past, put its social vision into a physical context. New towns, neighbourhood units, out-county estates, green belts, national parks, urban-regeneration programmes are all examples of planning concepts that have been motivated by social purpose.

A focus on sustainability and social purpose would be the contribution that planning brings to the quest for sustainable development. Achieving a sustainable society will require a much greater role for planning than currently seems probable. Sustainability requires an institutional and political setting that gives priority to long-term goals, in other words a system based on the precautionary approach, that reflects the integrated nature of environmental processes and that takes a strategic view of decision taking. Unfashionable (and, in many quarters, unthinkable) concepts such as intervention, collective action, socialism, communitarian values, the public interest and social equality will have to be revivified if we are to secure survival in the long term.

All this may sound hopelessly unrealistic at present. If, as seems highly probable, ecological modernisation, as a means of averting environmental crisis, proves to be an illusion, then alternatives can no longer be dismissed as fantasy. The risks of continuing as we are gradually become more evident. The task now is to peer further ahead, to seek to understand the processes of change and to work towards developing alternative, more sustainable and socially more equal ways of living in our environment. It is a task that planners, above all, must be capable of fulfilling.

BIBLIOGRAPHY

Abercrombie, N. and Urry, J. (1983) *Capital, Labour and the Middle Classes*, Allen & Unwin, London.

Addison, P. (1977) *The Road to 1945: British Politics and the Second World War*, Jonathan Cape, London.

Ambrose, P. (1986) *Whatever Happened to Planning?* Methuen, London.

Anson, B. (1981) *I'll Fight You for It! Behind the Struggle for Covent Garden*, Jonathan Cape, London.

Anton, T. J. (1975) *Governing Greater Stockholm: A Study of Policy Development and System Change*, California University Press, Berkeley.

Ashworth, G. and Voogd, H. (1990) *Selling the City: Marketing Approaches in Public Sector Urban Planning*, Belhaven, London.

Ashworth, W. (1954) *The Genesis of Modern British Town Planning: A Study in Economic and Social History of the Nineteenth and Twentieth Centuries*, Routledge, London.

Babtie Ltd (for Royal County of Berkshire) (1993) *Proof of Evidence: General Philosophy of the Plan A: The Council's Case*, Public Inquiry into the Draft Replacement Minerals Local Plan for Berkshire, Document BCC/2A, Reading, Babtie Public Services Division.

Ball, R. and Pratt, A. (1994) *Industrial Property: Policy and Economic Development*, Routledge, London.

Barnes, N. (1996) 'Conflicts over Biodiversity', in Sloep, P. and Blowers, A. (eds), *Environmental Policy in an International Context: 2. Environmental Problems as Conflicts of Interest*, Edward Arnold, London.

Barrett, S. and Fudge, C. (eds) (1981) *Policy and Action: Essays on the Implementation of Public Policy*, Methuen, London.

Bater, J. H. (1980) *The Soviet City: Ideal and Reality*, Edward Arnold, London.

Beck, U. (1992) *Risk Society: Towards a New Modernity*, Sage, London.

Beck, U. (1995) *Ecological Politics in an Age of Risk*, Polity Press, Cambridge.

Beck, U., Giddens, A. and Lash, S. (1994) *Reflexive Modernisation: Politics, Tradition and Aesthetics in the Modern Social Order*, Polity Press, Cambridge.

Bedfordshire County Council (1995) *Structure Plan 2011*, deposit draft, February.

Berkshire County Council (1975) *Reading, Wokingham, Aldershot, Basingstoke Subregional Study: Report of the Study Team*, Berkshire County Council, Reading.

Blowers, A. (ed.) (1993a) *Planning for a Sustainable Environment*, A Report by the Town and Country Planning Association, Earthscan, London.

Blowers, A. (1993b) 'Environmental Policy: The Quest for Sustainable Development', *Urban Studies*, 30 (4/5), 775–96.

Blowers, A. and Leroy, P. (1994) 'Power, Politics and Environmental Inequality: A Theoretical and Empirical Analysis of the Process of "Peripheralisation"', *Environmental Politics*, 3 (2), 197–228.

Blunkett, D. and Jackson, K. (1987) *Democracy in Crisis: The Town Halls Respond*, Hogarth Press, London.
Boddy, M. and Fudge, C. (1984) *Local Socialism? Labour Councils and New Left Alternatives*, Macmillan, London.
Bramble, B. and Porter, G. (1992) 'Non-Governmental Organizations and the Making of US International Environmental Policy', in Hurrell, A. and Kingsbury, B. (eds), *The International Politics of the Environment*, Clarendon Press, Oxford.
Breheny, M. (ed.) (1992) *Sustainable Development and Urban Form*, Pion, London.
Breheny, M. and Bately, P. (1982) *The History of Planning Methodology: A Framework for the Assessment of Anglo-American Theory and Practice*, Geographical Papers, Department of Geography, University of Reading.
Breheny, M. and Congden, P. (eds) (1989) *Growth and Change in a Core Region: The Case of South East England*, Pion, London.
Breheny, M. and Hall, P. (eds) (1996) *The People – Where Will They Go? National Report of the TCPA Regional Inquiry into Housing Need and Provision in England*, Town and Country Planning Association, London.
Brindley, T., Rydin, Y. and Stoker, G. (1989) *Remaking Planning: The Politics of Urban Change in the Thatcher Years*, Unwin Hyman, London.
Bruce, M. (ed.) (1973) *The Rise of the Welfare State: English Social Policy, 1601–1971*, Weidenfeld & Nicolson, London.
Brundell, M. J. (1994) *Inspector's Report of the Inquiry into the Replacement Minerals Local Plan for Berkshire* (21 September to 16 November 1993), Planning Inspectorate, Bristol.
Brundtland, G. H. (1987) *Our Common Future: Report of the World Commission on Environment and Development*, Oxford University Press, Oxford.
Bunge, W. (1971) *Geography of a Revolution*, Schenkman Publishing, Cambridge, Mass.
Cairncross, F. (1995) *Green Inc.: A Guide to Business and the Environment*, Earthscan, London.
Calder, A. (1968) *The People's War*, Jonathan Cape, London.
Castles, F. G. (1978) *The Social Democratic Image of Society: A Study of the Achievements and Origins of Scandinavian Social Democracy in Comparative Perspective*, Routledge & Kegan Paul, London.
CEC (Commission of the European Communities) (1990) *Green Paper on the Urban Environment* (EUR 12902 EN), Commission of the European Communities, Brussels.
CEC (1992) *Towards Sustainability: A New European Programme of Policy and Action in Relation to the Environment and Sustainable Development*, Com (92) 23 Final, CEC, Brussels.
Chambers, R. (1986) 'Putting the Last First', in Ekins, P. (ed.), *The Living Economy*, Routledge, London.
Chambers, R. (1993) *Challenging the Professions*, Intermediate Technology Publications, London.
Cherry, G. (1974) *The Evolution of British Town Planning: A History of Town Planning in the United Kingdom During the Twentieth Century, and of the Royal Town Planning Institute 1914-1974*, Leonard Hill, Leighton Buzzard.
Cherry, G. (1981) *The Politics of Town Planning*, Longman, Harlow.
Collis, I., Jacobs, M. and Heap, J. (1992) *Strategic Planning and Sustainable Development*, paper prepared for English Nature, English Nature, Peterborough.
Community Action (1972) 'Bureaucratic Guerrilla: An Alternative Role for Local Authority Planners', *Community Action*, February.
Cooke, P. (1990) *Back to the Future*, Unwin Hyman, London.
Cordy, T. (1996) Editorial, *Town and Country Planning*, 65 (2), 35.
Countryside Commission, English Heritage and English Nature (1993) *Conservation*

Issues in Strategic Plans, CCP 420, The Countryside Commission, Cheltenham.

County Planning Officers' Society (1993) *Planning for Sustainability*, CPOS, available from Peter Bell, Hampshire County Council, Winchester.

Cowell, R. (1993) *Take and Give: Managing the Impacts of Development with Environmental Compensation*, UKCEED Discussion Paper No. 10, UK Centre for Economic and Environmental Development, Cambridge.

Cox, A. (1984) *Adversary Politics and Land: The Conflict over Land and Property Policy in Post-War Britain*, Cambridge University Press, Cambridge.

Crawley, I. (1992) '... Or Consultation by Those who Must Implement the Plan?', *Town and Country Planning*, 61 (10), 275.

Critchley, J. (1994) *Heseltine* (revised edition), André Deutsch, London.

Crompton, R. (1990) 'Professions in the Current Context', *Work, Employment and Society*, British Sociological Association, London, pp. 147–66.

Crosland, S. (1982) *Tony Crosland*, Jonathan Cape, London.

CSERGE (Centre for Social and Economic Research on the Global Environment) (1993) 'Sustainable Development in the United Kingdom', in Green College Centre for Environmental Policy and Understanding (ed.), *Proceedings of Sustainable Development Seminar*, Green College, Oxford.

Cullingworth, J. B. (1975) *Environmental Planning, 1939–1969. Vol. 1: Reconstruction and Land Use Planning, 1939–1947* [*Peacetime History*], HMSO, London.

Cullingworth, J. B. (1979) *Environmental Planning, 1939–1969. Vol. 3: New Towns Policy* [*Peacetime History*], HMSO, London.

Cullingworth, J. B. (1980) *Environmental Planning, 1939-1969. Vol. 4: Land Values, Compensation and Betterment* [*Peacetime History*], HMSO, London.

Cullingworth, J. B. and Nadin, V. (1994) *Town and Country Planning in Britain*, Routledge, London.

Cutter, S. (1993) *Living with Risk*, Edward Arnold, London.

Daly, H. (1992) *Steady-state Economics*, Earthscan, London.

Daly, H. E. and Cobb Jr., J. (1989) *For the Common Good*, Merlin Press, London.

Davies, J. G. (1972) *The Evangelistic Bureaucrat*, Tavistock, London.

Dennis, N. (1972) *Public Participation and Planners' Blight*, Faber, London.

de Swaan, A. (1988) *In Care of the State*, Polity Press, Cambridge.

Dickens, P., Duncan, S., Goodwin, M. and Gray, F. (1985) *Housing, States and Localities*, Methuen, London.

DOE (Department of the Environment) (1972) *New Local Authorities. Management and Structure*, HMSO, London.

DOE (1973) *Greater London Development Plan: Report of the Panel of Inquiry. Vol. 1: Report*, HMSO, London.

DOE (1977a) *Inner Area Studies: Liverpool, Birmingham, and Lambeth: Summary of Consultants' Final Reports*, HMSO, London.

DOE (1977b) *Inner London: Proposals for Dispersal and Balance: Final Report of the Lambeth Inner Area Study*, HMSO, London.

DOE (1977c) *Unequal City: Final Report of the Birmingham Inner Area Study*, HMSO, London.

DOE (1977d) *Change or Decay: Final Report of the Liverpool Inner Area Study*, HMSO, London.

DOE (1977e) *Policy for the Inner Cities*, HMSO,London.

DOE (1978) *Strategic Plan for the South East: Review: Government Statement*, HMSO, London.

DOE (1980) *Development Control: Policy and Practice*, Circular 22/80, HMSO, London.

DOE (1983) *Streamlining the Cities: Government Proposals for Reorganising Local Government in Greater London and the Metropolitan Counties* (Cmd 0062), HMSO, London.

DOE (1989) *Minerals Planning Guidance: Guidelines for Aggregates Provision in England*, MPG 6, HMSO, London. (Now superseded by DOE 1994b.)

DOE (1990) *This Common Inheritance: Britain's Environmental Strategy* (Cmnd1200), HMSO, London.

DOE (1992a) *Planning Policy Guidance: Housing* (PPG 3 (revised)), HMSO, London.

DOE (1992b) *Development Plans and Regional Planning Guidance* (PPG12), HMSO, London.

DOE (1992c) *General Policy and Principles*, Planning Policy Guidance note 1 (PPG1), HMSO, London.

DOE (1993) *Guidelines for Aggregates Provision in England and Wales*, Revision of MGP6, draft consultation document.

DOE (1994a) *Minerals Planning Guidance: Guidelines for Aggregates Provision in England* (MPG 6), HMSO, London.

DOE (1994b) *Planning Policy Guidance: Nature Conservation* (PPG 9), HMSO, London.

DOE (1994c) *Quality in Town and Country*, Department of the Environment, London.

DOE (1995) *Projections of Households in England to 2016*, HMSO, London.

DOE and Department of Transport (1993) *Reducing Transport Emissions through Planning* (ECOTEC Research and Consulting Ltd in Association with Transportation Planning Associates), HMSO, London.

DOE and Department of Transport (1994) *Planning Policy Guidance: Transport* (PPG 13), HMSO, London.

DOE and Welsh Office (1986) *The Future of Development Plans: A Consultation Paper*, Department of the Environment, London.

DOE and Welsh Office (1993) *Planning Policy Guidance: Town Centres and Retail Developments* (PPG 6 (revised)), HMSO, London.

DOE and Welsh Office (1995) *Planning Policy Guidance: Town Centres and Retail Developments* (PPG 6 (revised)), HMSO, London.

Dobson, A. (ed.) (1991) *The Green Reader*, André Deutsch, London.

Dobson, A. (1995a) 'No Environmentalisation Without Democratisation', *Town and Country Planning*, 64 (12), 322–3.

Dobson, A. (1995b) *Green Political Thought*, Routledge, London.

Donnison, D. V. and Soto, P. (1980) *The Good City: A Study of Urban Development and Policy in Britain*, Heinemann, London.

Douglas, R. (1976) *Land, People and Politics: A History of the Land Question in the United Kingdom, 1878–1952*, Allison & Busby, London.

Douglas, R. (1993) *Land and Labour*, Labour Land Campaign in association with the Land Value Taxation Campaign, London.

Dovkants, K. and Bar-Hillel, M. (1992) 'Ghost City, 1992', *London Evening Standard*, 29 May, pp. 12–13.

Dubash, N. and Oppenheimer, M. (1992) 'Modifying the Mandate of Existing Institutions: NGOs', in Mintzer, I. (ed.), *Confronting Climate Change: Risks, Implications and Responses*, Cambridge University Press, Cambridge.

Duncan, S. S. (1985) 'Land Policy in Sweden: Separating Ownership from Development', in Barrett, S. M. and Healey, P. (eds), *Land Policy: Problems and Alternatives*, Gower, Aldershot.

Eckersley, R. (1992) *Environmentalism and Political Theory*, UCL Press, London.

Elkin, T., McLaren, D. and Hillman, M. (1991) *Reviving the City: Towards Sustainable Urban Development*, Friends of the Earth/Policy Studies Institute, London.

Engels, F. (1959) 'Socialism: Utopian and Scientific', in Feuer, L. S. (ed.), *Marx and*

Engels: Basic Writings on Politics and Philosophy, Doubleday Anchor, New York, pp. 68–111.

English Nature (1994) *Planning for Environmental Sustainability*, report prepared for English Nature by David Tyldesley and Associates, English Nature, Peterborough.

Evans, B. (1993) 'Why We no Longer Need a Planning Profession', *Planning Practice and Research*, 8 (1), 9–15.

Evans, B. (1995a) 'Can we have Sustainability without the Recoupment of Development Value?', in Elworthy, S., Anderson, K., Coates, I., Stephens, P. and Stroh, M. (eds), *Perspectives on the Environment*, Avebury, Aldershot.

Evans, B. (1995b) *Experts and Environmental Planning*, Avebury, Aldershot.

Faisalabad Development Authority (1986) *Structure Plan of Faisalabad*, Faisalabad Development Authority, Faisalabad.

Fischer, F. and Forester, J. (1993) *The Argumentative Turn in Policy Analysis and Planning*, UCL Press, London.

Freidson, E. (1984) 'Are Professions Necessary?', in Haskell, J. (ed.), *The Authority of Experts*, Indiana University Press, Bloomington, pp. 3–27.

Gare, M. and Johnson, T. (1993) *Foucault's New Domains*, Routledge, London.

George, S. (1988) *A Fate Worse Than Debt*, Penguin, London.

Gibbons, M. *et al.* (1994) *The New Production of Knowledge*, Sage, London.

Giddens, A. (1990) *The Consequences of Modernity*, Polity Press, Cambridge.

Government Office for London (1995) *Thames Strategy: A Study of the Thames Prepared for the Government Office for London*, HMSO, London.

Griffiths, R. (1990) 'Planning in Retreat? Town Planning and the Market in the 1980s', in Montgomery, J. and Thornley, A. (eds), *Radical Planning Initiatives: New Directions for Urban Planning in the 1990s*, Gower, Aldershot.

Hague, C. (1984) *The Development of Planning Thought: A Critical Perspective*, Hutchinson, London.

Hague, C. (1990) 'Planning and Equity in Eastern Europe: Raking Through the Rubble', *The Planner*, 30 March, pp. 19–21.

Hall, P. (ed.) (1965) *Land Values*, Sweet & Maxwell, London.

Hall, P. (1989) *London 2001*, Unwin Hyman, London.

Hall, P., Thomas R., Gracey, G. and Drewett, R. (1973) *The Containment of Urban England* (2 vols), Allen & Unwin, London.

Hall, S. (1988) *The Hard Road to Renewal: Thatcherism and the Crisis of the Left*, Verso, London.

Hallett, G. (1977) *Housing and Land Policies in West Germany and Britain: A Record of Success and Failure*, Macmillan, London.

Hallett, G. (1979) *Urban Land Economics: Principles and Policy*, Macmillan, London.

Halsey, A. H. (1987) 'Social Trends Since World War II', *Social Trends*, HMSO, London.

Hamer, M. (1988) *Wheels Within Wheels: A Study of the Road Lobby*, Routledge & Kegan Paul, London.

Hammersley, R., Maier, K. and Westlake, T. (1994) 'Czechs Fight Legacy of Total Planning Era', *Planning*, 1051, 14 January, pp. 1051–2.

Hansen, A. (ed.) (1993) *The Mass Media and Environmental Issues*, Leicester University Press, Leicester.

Hardin, G. (1968) 'The Tragedy of the Commons', *Science*, 162, 1243–8.

Harvey, D. (1990) *The Condition of Postmodernity*, Blackwell, Oxford.

Haskell, T. (ed.) (1984) *The Authority of Experts*, Indiana University Press, Bloomington.

Hass-Klau, C. (1990) *The Pedestrian and City Traffic*, Belhaven, London.

Healey, P. *et al.* (eds) (1992) *Rebuilding the City: Property-led Urban Regeneration*, E. & F. N. Spon, London.

Healey, P. (1993) 'The Communicative Work of Development Plans', *Environment and Planning* B, 20, 83–194.

Healey, P. (1995) 'Discourse of Integration: Making Frameworks for Democratic Urban Planning', in Healey, P. *et al.*, *Managing Cities: The New Urban Context*, Wiley, London, pp. 251–72.

Healey, P. (1997) *Collaborative Planning*, Macmillan, London.

Healey, P., McDougall, G. and Thomas, M. J. (1982) *Planning Theory: Prospects for the 1980s*, Pergamon, Oxford.

Healey, P. and Shaw, T. (1994a) 'Changing Meanings of "Environment" in the British Planning System', *Transactions of the Institute of British Geographers*, 19 (4), 425–38.

Healey, P. and Shaw, T. (1994b) *The Treatment of Environment by Planners: Evolving Concepts and Policies in Development Plans*, Working Paper no. 31, Department of Town and Country Planning, University of Newcastle upon Tyne.

Healey, P. and Underwood, J. (1978) 'Professional Ideals and Planning Practice: A Report on Research into Planners' Ideas in Practice in London Borough Planning Departments', *Progress in Planning*, 9, 73–127.

Heclo, H. and Madsen, M. (1987) *Policy and Politics in Sweden: Principled Pragmatism*, Temple University Press, Philadelphia.

Hennessy, P. (1992) *Never Again: Britain 1945–1951*, Jonathan Cape, London.

Heskin, A. D. (1992) 'Ethnicity, Race, Class, and Ideology Come Together in LA', in Smith, M. (ed.), *After Modernism: Global Restructuring and the Changing Boundaries of City Life*, Transactions Publishers, New Jersey.

Hillman, M., Adams, J. and Whitelegg, J. (1990) *One False Move: A Study of Children's Independent Mobility*, Policy Studies Institute, London.

Hirst, P. (1989) *After Thatcher*, Collins, London.

Hirst, P. and Thompson, G. (1996) *Globalisation in Question*, Polity Press, Cambridge.

HMSO (1991) *Policy Appraisal and the Environment: Environmental Appraisal in Government Departments*, London.

HMSO (1994a) *Sustainable Development: The UK Strategy*, Cm 2426, London.

HMSO (1994b) *Climate Change: The UK Programme*, Cm 2427M, London.

HMSO (1994c) *Biodiversity: The UK Action Plan*, Cm 2428, London.

Hollingsworth, M. (1986) *The Press and Political Dissent*, Pluto Press, London.

Holmberg, J., Bass, S. and Timberlake, L. (1991) *Defending the Future*, Earthscan, London.

Holmberg, J., Bass, S. and Timberlake, L. (1993) *Facing the Future*, Earthscan, London.

House of Lords Select Committee on Sustainable Development (1995) *Report on Sustainable Development*, HL Paper 72, HMSO, London.

Howard, E. (1902) *Garden Cities of Tomorrow*, Swan Sonnenschein, London.

Hutton, W. (1995) *The State We're In*, Jonathan Cape, London.

Illich, I. (1977) *Disabling Professions*, Marion Boyars, London.

International Broadcasting Trust (1994) *The War Machine* (Study guide to series of television programmes on the international arms industry), Yorkshire Television, London.

Jacobs, M. (1991) *The Green Economy*, Pluto Press, London.

Jacobs, M. (1993) *Sense and Sustainability*, Council for the Protection of Rural England, London.

Jacobs, M. and Stott, M. (1992) 'Sustainable Development and the Local Economy', *Local Economy* 7 (3), 261–72.

James, G. (1995) *In the Public Interest*, Little, Brown, London.

Jobse, R. B. and Musterd, S. (1992) 'Changes in the Residential Function of the Big Cities', in Dieleman, F. M. and Musterd, S. (eds), *The Randstad: A Research and*

Policy Laboratory, Kluwer Academic Publishers, Dordrecht.
Johnson, T. J. (1972) *Professions and Power*, Macmillan, Basingstoke.
Johnson, T. J. (1993) 'Expertise and the State', in Gare, M. and Johnson, T., *Foucault's New Domains*, Routledge, London, pp. 139–52.
Kansky, K. J. (1976) *Urbanization under Socialism: The Case of Czechoslovakia*, Praeger, New York.
Keeble, L. (1952) *Principles and Practice of Town and Country Planning*, The Estates Gazette, London.
Kent Thames-Side Development Agency (1995) *Kent Thames-Side*, Kent Thames-Side Development Agency, Dartford.
Khakee, A., Elander, I. and Sunesson, S. (eds) (1995) *Remaking the Welfare State: Urban Planning and Policy-making in Sweden in the 1990s*, Avebury, Aldershot.
Kingdom, J. (1992) *No Such Thing as Society? Individualism and Community*, Open University Press, Buckingham.
Krishnarayan, V. and Thomas, H. (1993) *Ethnic Minorities and the Planning System*, Royal Town Planning Institute, London.
Labour Party, The (1994) *In Trust for Tomorrow*, London.
Lang, T. and Hines, C. (1993) *The New Protectionism*, Earthscan, London.
Larson, M. S. (1977) *The Rise of Professionalism: A Sociological Analysis*, University of California Press, Berkeley.
Larson, M. S. (1984) 'The Production of Expertise and the Constitution of Expert Power', in Haskell, T. (ed.), *The Authority of Experts*, Indiana University Press, Bloomington, pp. 28-80.
Larson, M. S. (1990) 'In the Matter of Experts and Professionals, or How Impossible it is to Leave Nothing Unsaid', in Torstendahl, R. and Burrage, M. (eds), *The Formation of Professions*, Swedish Collegium for Advanced Study, Stockholm, pp. 24–50.
Lash, S. and Urry, J. (1987) *The End of Organised Capitalism*, Polity Press, Cambridge.
Levett, R. (1993) *Agenda 21: A Guide For Local Authorities in the UK, Local Government Management Board*, LGMB, Luton.
LGMB (Local Government Management Board) (1993) *Local Agenda 21 in the UK*, LGMB, Luton.
Loeb, P. (1986) *Nuclear Culture*, New Society Publishers, Philadelphia.
Lovelock, J. (1989) *The Ages of Gaia*, Oxford University Press, Oxford.
Lowe, M. D. (1992) 'Shaping Cities', in Brown, L. R. and others, *State of the World, 1992: A Worldwatch Institute Report on Progress Toward a Sustainable Society*, Earthscan, London, pp. 119–37.
Luper-Foy, S. (1992) 'Justice and Natural Resources', *Environmental Values*, 1, 47–64.
Mabey, R. (1993) 'Analysis of Responses to the Draft MPG6', unpublished paper, Minerals Planning Division, Department of the Environment, London.
Macdonald, K. (1995) *The Sociology of the Professions*, Sage, London.
MacEwen, M. and MacEwen, A. (1982) *National Parks: Conservation or Cosmetics?* Allen & Unwin, London.
McGrew, A. (1992), 'Conceptualizing Global Politics', in McGrew, A. G. and Lewis, P. G., *Global Politics*, Polity Press, Cambridge.
McKay, D. H. and Cox, A. W. (1979) *The Politics of Urban Change*, Croom Helm, London.
McLoughlin, J. B. (1969) *Urban and Regional Planning: A Systems Approach*, Faber, London.
McSorley, J. (1990) *Living in the Shadow*, Pan Books, London.
Maier, K. (1994) 'Planning and an Education in Planning in the Czech Republic', *Journal of Planning Education and Research*, 13, 263–9.

Marsh, D. and Rhodes, R. A. W. (eds) (1992) *Implementing Thatcherite Policies: Audit of an Era*, Open University Press, Buckingham.
Massey, D. (1984) *Spatial Divisions of Labour: Social Structure and the Geography of Production*, Macmillan, London.
Massey, D. and Allen, J. (1988) *The Economy in Question*, Sage, London.
Massey, D. and Catalano, A. (1978) *Capital and Land*, Edward Arnold, London.
Massey, D, Quintas, P. and Wield, D. (1992) *High Tech Fantasies*, Routledge, London.
Mazza, K. and Rydin, R. (eds) (1996) 'Urban Sustainability: Actors, Discourses, Networks and Policy Tools', *Progress in Planning*, 47(1), 1–74.
Meadows, D. M., Meadows, D., Randers, J. and Behrens, W. W. (1972) *The Limits to Growth*, Universe Books, New York.
Meadows, D. M., Meadows, D. L. and Randers, J. (1992) *Beyond the Limits*, Earthscan, London.
Meegan, R. (1988) 'A Crisis of Mass Production?', in Massey, D. and Allen, J. (eds), *The Economy in Question*, Sage, London.
Minister without Portfolio (1985) *Lifting the Burden: Presented to Parliament by the Minister without Portfolio* (Cmnd. 9571), HMSO, London.
Ministry of Transport (1970) *Transport Planning: The Men for the Job: A Report to the Minister of Transport by Lady Sharp*, HMSO, London.
Mol, A. (1995) *The Refinement of Production: Ecological Modernization Theory and the Chemical Industry*, Van Arkel, Utrecht.
Montgomery, J. and Thornley, A. (eds) (1990) *Radical Planning Initiatives: New Directions for Urban Planning in the 1990s*, Gower, Aldershot.
Montin, S. and Elander, I. (1995) 'Citizenship, Consumerism and Local Government in Sweden', *Scandinavian Political Studies*, 18 (1), 25–51.
Myerson, G. and Rydin, Y. (1994) '"Environment" and Planning: A Tale of the Mundane and the Sublime', *Society and Space*, 12, 437–52.
Myerson, G. and Rydin, Y. (1996a) 'Sustainable Development: The Implications of the Global Debate for Land Use Planning', in Buckingham-Hatfield, S. and Evans, B. (eds), *Environmental Planning and Sustainability*, Wiley, Chichester.
Myerson, G. and Rydin, Y. (1996b) *The Language of Environment: A New Rhetoric*, UCL Press, London.
NEPP (National Environmental Policy Plan) (1989) *To Choose or to Lose*, Ministry of Housing, Physical Planning and Environment Department for Information and International Relations, The Hague, The Netherlands.
O'Riordan, T. (1992) 'The Environment', in Cloke, P. (ed.), *Policy and Change in Thatcher's Britain*, Pergamon, Oxford.
Owens, S. (1991) *Energy-conscious Planning*, Campaign for the Protection of Rural England, London.
Owens, S. E. (1994) 'Land, Limits and Sustainability: A Conceptual Framework and Some Dilemmas for the Planning System', *Transactions of the Institute of British Geographers*, 19 (4), 439–56.
Pearce, D. W. and Turner, R. K. (1990) *Economics of Natural Environments and the Environment*, Harvester Wheatsheaf, Hemel Hempstead.
Pearce, D., Barbier, E. and Markandya, A. (1990) *Sustainable Development: Economics and Environment in the Third World*, Earthscan, London.
Pearce, D. W., Markandya, A. and Barbier, E. (1989) *Blueprint for a Green Economy*, Earthscan, London.
Pepper, D. (1993) *Eco-Socialism: From Deep Ecology to Social Justice*, Routledge, London.
Peters, T. (1989), *Thriving on Chaos: A Handbook for a Management Revolution*, Ferozsons, Lahore.
Pezzey, J. (1989) *Definitions of Sustainability*, UKCEED Discussion Paper No. 9, UK

Centre for Economic and Environmental Development, Cambridge.
Porritt, J. (1996) 'Real World Politics', *Town and Country Planning*, 65 (2), 34.
Potter, D. (1995) 'Environmental Problems in their Political Context', in Glasbergen, P. and Blowers, A. (eds), *Environmental Policy in an International Context: Vol. 1: Perspectives on Environmental Problems*, Edward Arnold, London.
Raemakers, J. (ed.) (1992) *Local Authority Green Plans: A Practical Guide*, Working Paper No. 39, Department of Planning, Heriot Watt University.
Ravetz, A. (1980) *Remaking Cities: Contradications of Recent Urban Development*, Croom Helm, London.
Ravetz, A. (1986) *The Government of Space: Town Planning in Modern Society*, Faber, London.
Reade, E. J. (1987) *British Town and Country Planning*, Open University Press, Milton Keynes.
Reade, E. J. (1989a) English summary of Asel Floderus, *The Credibility and Legitimacy of Physical Planning*, The Swedish National Institute for Building Research, Gävle, Sweden.
Reade, E. J. (1989b) Editor's preface to research report, *Britain and Sweden: Current Issues in Local Government*, The Swedish National Institute for Building Research, Gävle, Sweden.
Reade, E. J. (1994) 'Openness in Development Control', unpublished discussion paper.
Redclift, M. (1990) *Sustainable Development: Exploring the Contradictions*, Methuen, London.
Rhodes, R. (1994) 'The Hollowing Out of the State: The Changing Nature of the Public Service in Britain', *The Political Quarterly*, 65 (2), 138–51.
RIVM (1989) *Concern for Tomorrow*, National Institute of Public Health and Environmental Protection, Bilthoven, The Netherlands.
Robertson, D. (1995) 'United Reforms', *Town and Country Planning*, 64 (10), 278–9.
Royal Commission on Local Government in England (1969) *Report* (Cmnd. 4040), HMSO, London.
RCB (Royal County of Berkshire) (1993) *Draft Replacement Minerals Local Plan for Berkshire*, RCB, Reading.
RSPB (Royal Society for the Protection of Birds) (1993) *The UK Biodiversity and Sustainability Plans: Comments on their Preparation and Basis*, RSPB, Sandy, Bedfordshire.
Rydin, Y. (1993) *The British Planning System*, Macmillan, London.
Rydin, Y. (1994) 'Sustainable Development, Equity Considerations and the Role of Land Use Planning', mimeo.
Rydin, Y. (1995) 'Sustainable Development and the Role of Land Use Planning', *Area*, 27 (4), 369–77.
Rydin, Y. and Greig, A. (1994) 'Talking Past Each Other: Environmentalists in Different Organisational Settings', mimeo.
Secretary of State for Trade and Industry (1988) *Releasing Enterprise: Presented to Parliament by the Secretary of State for Trade and Industry* (Cm. 512), HMSO, London.
SERPLAN (South-East Regional Planning Council) (1989) *The Apportionment of the Production of Construction Aggregates in the South East up to 2006*, RPC 14461989, SERPLAN, London.
Siddiq, M. (1994) 'Evaluation of Approved Private Sites and Services Schemes in Faisalabad', unpublished MSc thesis, Department of City and Regional Planning, University of Engineering and Technology, Lahore.
Silvertown, J. (1990), 'Inhabitants of the Biosphere', in Silvertown, J. and Sarre, P. (eds), *Environment and Society*, Hodder & Stoughton, London, pp. 88–120.
Simmie, J. (1981) *Power, Property and Corporatism*, Macmillan, London.

Simmie, J. (1990) *Planning at the Crossroads*, UCL Press, London.

Skeffington Committee (1968) *People and Planning: Report of the Committtee on Public Participation in Planning*, HMSO, London.

Skidelsky, R. (ed.) (1988) *Thatcherism*, Chatto & Windus, London.

Smyth, G., Jones, D. and Platt, S. (eds) (1994) 'Bite the Ballot: 2500 Years of Democracy', *New Statesman and Society*, 29 April.

Soja, E. and Scott, A. (1986) 'Editorial: Los Angeles: Capital of the Late Twentieth Century', *Environment and Planning D: Society and Space*, 4 (3), 249–54.

South East Joint Planning Team (1970) *Strategic Plan for the South East: Report by the South East Joint Planning Team*, HMSO, London.

South East Joint Planning Team (1976) *Strategy for the South East: 1976 Review: Report with Recommendations by the South East Joint Planning Team*. HMSO, London.

Svensson, L. (1990) 'Knowledge as a Professional Resource: Case Studies of Architects and Psychologists at Work', in Torstendahl, R. and Burrage, M. (eds), *The Formation of Professions*, Swedish Collegium for Advanced Study, Stockholm, pp. 50–70.

Szeleyni, I. (1983) *Urban Inequalities Under State Socialism*, Oxford University Press, Oxford.

Thames Gateway Task Force (1995) *The Thames Gateway Planning Framework* (RPG 9a), Department of the Environment, London.

Thomas, C. (1993) 'Beyond UNCED: An Introduction', *Environmental Politics*, 2 (4), 1–27.

Thornley, A. (1991) *Urban Planning Under Thatcherism: The Challenge of the Market*, Routledge, London.

Thornley, A. (ed.) (1992) *The Crisis of London*, Routledge, London.

Tolley, R. (ed.) (1990) *The Greening of Urban Transport: Planning for Walking and Cycling in Western Cities*, Belhaven, London.

Torstendahl, R. and Burrage, M. (eds) (1990) *The Formation of Professions*, Swedish Collegium for Advanced Study, Stockholm.

Travers Morgan Engineering (1973) *Development Options for Dockland*, HMSO, London.

Turba, M. (1993) *The Physical, Ecological, Demographic and Social Barriers to the Future Development and Growth of Prague*, City Architect's Office of Prague, Prague.

Tyme, J. (1978) *Motorways versus Democracy: Public Inquiries into Road Proposals and Their Political Significance*, Macmillan, London.

United Nations (1991) *World Urbanization Prospects 1990: Estimates and Projections of Urban and Rural Populations and of Urban Agglomerations*, United Nations, ST/ESA/SER.A/121, New York.

United Nations (1992) *Agenda 21: The UN Programme of Action from Rio*, United Nations, New York.

United Nations Centre for Human Settlements (1990) *Global Strategy for Shelter to the year 2000*, Habitat, Nairobi.

Utthwatt Committee (1942) *Final Report of the Expert Committee on Compensation and Betterment*, HMSO, London.

Ward, S. (1993) 'Thinking Global, Acting Local? British Local Authorities and their Environmental Plans', *Environmental Politics*, 2 (3), 453–78.

Ward, S. V. (1994) *Planning and Urban Change*, Paul Chapman, London.

WCED (World Commission for Environment and Development) (1987) *Our Common Future*, Oxford University Press, Oxford.

Weale, A. (1992) *The New Politics of Pollution*, Manchester University Press, Manchester.

Welbank, M. (1993) *Sustainable Development: A Discussion Paper*, Royal Town Planning Institute, London.

Wilding, P. (1982) *Professional Power and Social Welfare*, Routledge, London.
Wynne, B. (1994) 'Scientific Knowledge and the Global Environment', in Redclift, M. and Benton, T. (eds), *Social Theory and the Global Environment*, Routledge, London.
Wynne, B. *et al.* (1993) *Public Perceptions and the Nuclear Industry in West Cumbria*, The Centre for the Study of Environmental Change, University of Lancaster.
Zonabend, F. (1989) *The Nuclear Peninsula*, Cambridge University Press, Cambridge.

INDEX